Improving Flood Detection and Monitoring through Remote Sensing

Improving Flood Detection and Monitoring through Remote Sensing

Editors

Alberto Refice
Domenico Capolongo
Marco Chini
Annarita D'Addabbo

MDPI • Basel • Beijing • Wuhan • Barcelona • Belgrade • Manchester • Tokyo • Cluj • Tianjin

Editors
Alberto Refice
National Research
Council—Institute for
Electromagnetic Sensing of
the Environment (CNR-IREA)
Italy

Domenico Capolongo
University of Bari
Italy

Marco Chini
Luxembourg Institute of
Science and Technology
(LIST)
Luxembourg

Annarita D'Addabbo
National Research
Council—Institute for
Electromagnetic Sensing of
the Environment (CNR-IREA)
Italy

Editorial Office
MDPI
St. Alban-Anlage 66
4052 Basel, Switzerland

This is a reprint of articles from the Special Issue published online in the open access journal *Water* (ISSN 2073-4441) (available at: https://www.mdpi.com/journal/water/special_issues/Flood_Remote_Sensing).

For citation purposes, cite each article independently as indicated on the article page online and as indicated below:

LastName, A.A.; LastName, B.B.; LastName, C.C. Article Title. *Journal Name* **Year**, *Volume Number*, Page Range.

ISBN 978-3-0365-3875-4 (Hbk)
ISBN 978-3-0365-3876-1 (PDF)

© 2022 by the authors. Articles in this book are Open Access and distributed under the Creative Commons Attribution (CC BY) license, which allows users to download, copy and build upon published articles, as long as the author and publisher are properly credited, which ensures maximum dissemination and a wider impact of our publications.

The book as a whole is distributed by MDPI under the terms and conditions of the Creative Commons license CC BY-NC-ND.

Contents

About the Editors vii

Alberto Refice, Domenico Capolongo, Marco Chini and Annarita D'Addabbo
Improving Flood Detection and Monitoring through Remote Sensing
Reprinted from: *Water* **2022**, *14*, 364, doi:10.3390/w14030364 1

Takahiro Sayama, Koji Matsumoto, Yuji Kuwano and Kaoru Takara
Application of Backpack-Mounted Mobile Mapping System and Rainfall–Runoff–Inundation Model for Flash Flood Analysis
Reprinted from: *Water* **2019**, *11*, 963, doi:10.3390/w11050963 5

Xianwei Wang, Lingzhi Wang and Tianqiao Zhang
Geometry-Based Assessment of Levee Stability and Overtopping Using Airborne LiDAR Altimetry: A Case Study in the Pearl River Delta, Southern China
Reprinted from: *Water* **2020**, *12*, 403, doi:10.3390/w12020403 21

Giuseppe Ruzza, Luigi Guerriero, Gerardo Grelle, Francesco Maria Guadagno and Paola Revellino
Multi-Method Tracking of Monsoon Floods Using Sentinel-1 Imagery
Reprinted from: *Water* **2019**, *11*, 2289, doi:10.3390/w11112289 41

David C. Mason, John Bevington, Sarah L. Dance, Beatriz Revilla-Romero, Richard Smith, Sanita Vetra-Carvalho and Hannah L. Cloke
Improving Urban Flood Mapping by Merging Synthetic Aperture Radar-Derived Flood Footprints with Flood Hazard Maps
Reprinted from: *Water* **2021**, *13*, 1577, doi:10.3390/w13111577 59

Mahmoud Rajabi, Hossein Nahavandchi and Mostafa Hoseini
Evaluation of CYGNSS Observations for Flood Detection and Mapping during Sistan and Baluchestan Torrential Rain in 2020
Reprinted from: *Water* **2020**, *12*, 2047, doi:10.3390/w12072047 83

Viktoriya Tsyganskaya, Sandro Martinis and Philip Marzahn
Flood Monitoring in Vegetated Areas Using Multitemporal Sentinel-1 Data: Impact of Time Series Features
Reprinted from: *Water* **2019**, *11*, 1938, doi:10.3390/w11091938 99

Alberto Refice, Marina Zingaro, Annarita D'Addabbo and Marco Chini
Integrating C- and L-Band SAR Imagery for Detailed Flood Monitoring of Remote Vegetated Areas
Reprinted from: *Water* **2020**, *12*, 2745, doi:10.3390/w12102745 123

About the Editors

Alberto Refice

Alberto Refice is a Researcher at the Italian National Research Council (CNR)—Institute for Electromagnetic Sensing of the Environment (IREA). His main research interests concern advanced processing techniques for remotely sensed data, from synthetic aperture radars (SARs) and optical sensors. He is involved in several projects concerning remote sensing applications such as environmental hazard monitoring and management, geo-hydrological process monitoring and modeling and landscape evolution.

Domenico Capolongo

Domenico Capolongo is an Associate Professor in physical geography and geomorphology at the Earth and Geo-Environmental Sciences Department, University of Bari "Aldo Moro". His main research interests concern the application of remote sensing in investigating surface processes and hydrogeomophological hazard and risks. He is involved in several projects on geomorphological mapping, GIS and remote sensing applications to flood and landslide hazards.

Marco Chini

Marco Chini is a Senior Research and Technology Associate at the Luxembourg Science and Technology Institute (LIST). His research objectives have always focused on achieving a better understanding, characterization and monitoring of land surfaces and their changes. These research efforts have the aim to develop and apply methodologies and algorithms enabling the monitoring of different phenomena on a large scale using remote sensing data, both SAR and optical. The projects he is involved in, both fundamental and applied in nature, are in the fields of floodwater and natural disaster mapping, classification, soil moisture retrieval, geophysical parameter estimation and InSAR techniques applied to geophysical phenomena, e.g., earthquake and volcanic eruptions.

Annarita D'Addabbo

Annarita D'Addabbo is a Researcher at the Italian National Research Council (CNR)—Institute for Electromagnetic Sensing of the Environment (IREA). Her research activity is focused on the theoretical and experimental analysis of different classification methodologies, statistical models and machine learning techniques, applied to land cover change detection problems and to the monitoring of natural hazards from remote sensing data. She is involved in several national and international projects concerning the development of automatic tools devoted to environmental hazard detection and mapping or devoted to land use/cover mapping.

Editorial

Improving Flood Detection and Monitoring through Remote Sensing

Alberto Refice [1,*], Domenico Capolongo [2], Marco Chini [3] and Annarita D'Addabbo [1]

1. Consiglio Nazionale delle Ricerche – Istituto per il Rilevamento Elettromagnetico dell'Ambiente (CNR-IREA) 70126 Bari, Italy; annarita.daddabbo@cnr.it
2. Earth and Geoenvironmental Science Department, University of Bari, 70126 Bari, Italy; domenico.capolongo@uniba.it
3. Luxembourg Institute of Science and Technology (LIST), 4362 Esch-sur-Alzette, Luxembourg; marco.chini@list.lu
* Correspondence: alberto.refice@cnr.it

Citation: Refice, A.; Capolongo, D.; Chini, M.; D'Addabbo, A. Improving Flood Detection and Monitoring through Remote Sensing. *Water* **2022**, *14*, 364. https://doi.org/10.3390/w14030364

Received: 17 January 2022
Accepted: 20 January 2022
Published: 26 January 2022

Publisher's Note: MDPI stays neutral with regard to jurisdictional claims in published maps and institutional affiliations.

Copyright: © 2022 by the authors. Licensee MDPI, Basel, Switzerland. This article is an open access article distributed under the terms and conditions of the Creative Commons Attribution (CC BY) license (https://creativecommons.org/licenses/by/4.0/).

1. Introduction

Floods are among the most threatening and impacting environmental hazards. Their costs in terms of human lives, infrastructure damage or loss, and agricultural impact can be enormous and continue to increase due to climate change.

Investigating effects and extents of flood events in short times after occurrence is of utmost importance in order to quantify damage, organize rescue measures, determine insurance refunds, and calibrate prediction models for risk assessment and management. In the last years, remote sensing is proving to be a strong aid in this direction by providing large amounts of data of the Earth's surface at low to null costs. The increasing number of spacecraft and sensors available calls for the use of sophisticated procedures and algorithms to extract useful information from such large datasets. In [1], several examples of precise tools for investigating the effects of inundations were presented. Since then, improvements in technology, data availability, and processing power have occurred.

This Special Issue is a collection of six articles and one technical note, which provide a wide overview of recent advances in these fields. The papers deal with various aspects of flood monitoring by using diverse sensors such as backpack-mounted 3-D optical cameras, airborne LiDAR, GNSS reflectometry, and spaceborne synthetic aperture radar (SAR) data analysis from multiple sensors and wavelengths. Test sites are located in various parts of the world, including China, Japan, Philippines, Mozambique, Iran, UK, Greece, and Turkey. The volume represents, therefore, a useful survey of methods to improve the performance of techniques concerning remote sensing of floods in the mapping phase, including the assessment of post-disaster flood damage, integration of observed and predicted flood impacts, and evaluation of flood prevention measures such as levees.

2. Summary

The authors of [2] describe an experiment in retrieving field information about the effects of a flash flood that occurred in Japan in 2017 by using backpack-mounted equipment consisting of a series of optical cameras, a laser scanner, a GNSS receiver, and an IMU unit. They also realized a DEM of a strip of terrain around the water course after the event and compared it to one that was previously available. The resulting updated DEM was integrated with that coming from a flood propagation model run with parameters pertaining to the event, showing consistent improvements with respect to previous runs in terms of adherence to ground truth measurements. The authors also offer some interesting insight about issues and advantages encountered during the survey on the ground, showing how, even with ground-based equipment, remote and automated sensing by using sophisticated sensors brings a significant step forward in data collection campaigns. The results demonstrate, among other things, how sediment deposition plays an important part in flood propagation. This is in line with other studies in this direction [3].

In [4], high-resolution LiDAR data (16 points/m^2 on average) acquired from an airborne sensor are analyzed in depth to derive information about the condition and weathering state of levees along a section of river coastlines in Southeastern China. Levee geometric parameters, such as crown height, waterside, and landside slopes, and elevation transects are evaluated with respect to model flood heights for several return periods, resulting in scores assigned to each levee parcel to quantify its robustness against water overtopping. Such detailed scores improve upon previous classifications, mainly based on ground campaigns, which involved single parameters and sparse sampling. The great detail placed in the evaluation testifies the potential of remotely sensed data to extract fine-resolution information over relatively large areas with a fraction of the time-personnel effort necessary to perform field surveys.

Technical note [5] illustrates an example of application of three of the most diffused and well-known techniques for separating water and non-water areas in SAR images, namely the manual histogram "valley emphasis" thresholding method, Otsu's threshold determination, and K-means clustering. The latter algorithm is run here with two clusters. The three methods are tested on a series of Sentinel-1 images taken during a series of strong rain episodes that occurred between the end of 2018 and the beginning of 2019 over a region in the Philippines. Results are compared visually, as well as in terms of total detected flooded area. Although no ground truth was available to assess the methods' performances more quantitatively, the scatter of the results (relative differences in detected flooded areas reaching as much as 60% between manual and K-means) provides a fair idea of the care that should be placed in choosing algorithms and procedures to detect floodwaters in SAR imagery and the important role of reference data to assess algorithms quantitatively.

In [6], a detailed analysis is proposed for an approach to flood vulnerability mapping that makes use of high-resolution, SAR-derived, and model-derived flood hazard maps. Results are assessed by using aerial photos available for three test sites during two flood events in UK. Various combinations of remotely sensed and model information are compared. It was concluded that SAR flood maps improve detection performance especially when surface flooded areas are not necessarily due to river water inundation, while modeled flood maps obtained from gauge data are most useful for areas where the SAR sensor cannot "see" due to its side-looking geometry, such as streets in high-density urban centers. Synergetic use of both data types results in detection accuracies of up to 94%, with false positive rates as low as 9%, improving by several percent points performances obtained with the use of SAR data only. As discussed in the paper, although high-resolution SAR data are still mostly not free and open access (with the significant exception of the European Sentinel-1 constellation, and with several other SAR sensors collections foreseen to become available as open access in the next future), their use is crucial for complex sites such as urban areas, especially those characterized by relatively high building densities as in many European countries, where high-resolution digital surface models (DSM) are necessary in order to properly model hydrological dynamics. In such cases, current sensors such as Sentinel-1, despite their unparalleled temporal acquisition frequency that is showing promising results in several applications [7,8], may still have insufficient spatial resolution, while higher-resolution sensors such as the Italian COSMO-SkyMed and the German TerraSAR-X lack temporal frequency.

A potentially innovative data source for flood monitoring is investigated in [9]. GNSS reflectometry is a technique that uses microwave signals emitted by global positioning system constellations and collected by suitable receivers to gain information about (bistatic) reflectivity of the Earth's surface. NASA's CYGNSS mission is a constellation of microsatellites carrying receivers to exploit GNSS signals, which has proven to be useful in many land and ocean studies [10]. In this case, CYGNSS-collected reflectivity signals were used to map inundated areas during an exceptional precipitation event that occurred in January 2020 over a large basin in Southeastern Iran. Despite its relatively low resolution

(25×25 km^2), the high temporal acquisition frequency (average of about 7 h, reaching about 2 h as a maximum) makes it a very promising tool to improve the timeliness of flood survey maps over large areas.

In [11], an in-depth discussion is presented about the application of a methodology [12] to detect flood water both on open areas and under vegetation (indicated, respectively, as "temporary open water" and "temporary flooded vegetation" in the article) from time series of SAR images. Sentinel-1 data are analyzed over three test sites in Greece and Turkey that were affected by floods in March and April 2015 and June 2017, respectively. Identifying flooded vegetation, typically through the detection of the "double bounce" scattering behavior that causes significant backscatter increase with respect to non-flooded conditions, is considered one of the most difficult tasks in flood monitoring by SAR data. In the article, various combinations of the two available Sentinel-1 polarization channels, namely VV and VH, are considered as possible features to detect different terrain classes. The relative importance of single polarization channels and the difference, sum, and ratio of the two were evaluated by using statistical means, using information extracted from optical images as reference data, to identify the most relevant one(s) for the detection of each class. Results point to single VV polarization as the most efficient for open water flood identification, while the sum of VV + VH polarizations is recommended for detecting flooded vegetation, although small variations in performance for the different features appear from one test site to another, which seems to suggest that choosing one feature over the others may actually be a site-dependent task.

The task of discriminating terrain cover during floods is also considered in [13], where both C-band (Sentinel-1) and L-band (ALOS 2) data are exploited in order to extract information about a prolonged, large inundation, which affected a part of the Zambesi river basin in Mozambique, Africa, from December 2014 to April 2015. No useful optical data are available for this event, as is normally the case over equatorial sites that are mostly clouded for long periods of time. Nevertheless, the analysis of multi-temporal series, aided by the use of the CORINE land cover database, available at a resolution of 100 m over Africa and the synergy between the different wave penetration and backscattering characteristics of the two sensors allowed the derivation of an informative set of multitemporal maps of flood evolution, as well as an integrated, multi-sensor map discriminating various types of ground features and situations, such as flooded crops, grassland, and forest in addition to open water areas.

3. Conclusions

The papers in this Special Issue cover a broad spectrum of techniques and data analyses aimed at improving the performance of flood monitoring activities from remotely sensed data. Improvements come essentially from (i) availability of innovative data sources, such as backpack-mounted 3-D cameras or spaceborne GNSS reflectometry; (ii) precise assessment and validation of algorithm performance through comparison with independent data; (iii) integration of multiple data sources such as multi-frequency, multi-sensor, and multi-temporal satellite SAR data. Results of tests over sites located throughout the world demonstrate the great potential of such methods to bring significant innovation and increase algorithm precision in the field of inundation monitoring.

Author Contributions: Conceptualization, A.R., D.C., M.C. and A.D.; writing—original draft preparation, A.R.; writing—review and editing, A.R., D.C., M.C. and A.D. All authors have read and agreed to the published version of the manuscript.

Funding: This research received no external funding.

Acknowledgments: Gratitude is due to the editors of the journal, as well as to the authors who contributed with their articles to this Special Issue. Finally, special thanks goes to the anonymous reviewers, who have contributed efficiently to improve the quality of the articles.

Conflicts of Interest: The authors declare no conflict of interest.

References

1. Refice, A.; D'Addabbo, A.; Capolongo, D. (Eds.) *Flood Monitoring through Remote Sensing*; Springer Remote Sensing/Photogrammetry; Springer International Publishing: Cham, Switzerland, 2018, ISBN 978-3-319-63958-1.
2. Sayama, T.; Matsumoto, K.; Kuwano, Y.; Takara, K. Application of Backpack-Mounted Mobile Mapping System and Rainfall–Runoff–Inundation Model for Flash Flood Analysis. *Water* **2019**, *11*, 963. [CrossRef]
3. Zingaro, M.; Refice, A.; D'Addabbo, A.; Hostache, R.; Chini, M.; Capolongo, D. Experimental Application of Sediment Flow Connectivity Index (SCI) in Flood Monitoring. *Water* **2020**, *12*, 1857. [CrossRef]
4. Wang, X.; Wang, L.; Zhang, T. Geometry-Based Assessment of Levee Stability and Overtopping Using Airborne LiDAR Altimetry: A Case Study in the Pearl River Delta, Southern China. *Water* **2020**, *12*, 403. [CrossRef]
5. Ruzza, G.; Guerriero, L.; Grelle, G.; Guadagno, F.M.; Revellino, P. Multi-Method Tracking of Monsoon Floods Using Sentinel-1 Imagery. *Water* **2019**, *11*, 2289. [CrossRef]
6. Mason, D.C.; Bevington, J.; Dance, S.L.; Revilla-Romero, B.; Smith, R.; Vetra-Carvalho, S.; Cloke, H.L. Improving Urban Flood Mapping by Merging Synthetic Aperture Radar-Derived Flood Footprints with Flood Hazard Maps. *Water* **2021**, *13*, 1577. [CrossRef]
7. Chini, M.; Pelich, R.; Pulvirenti, L.; Pierdicca, N.; Hostache, R.; Matgen, P. Sentinel-1 InSAR Coherence to Detect Floodwater in Urban Areas: Houston and Hurricane Harvey as A Test Case. *Remote Sens.* **2019**, *11*, 107. [CrossRef]
8. Li, Y.; Martinis, S.; Wieland, M.; Schlaffer, S.; Natsuaki, R. Urban Flood Mapping Using SAR Intensity and Interferometric Coherence via Bayesian Network Fusion. *Remote Sens.* **2019**, *11*, 2231. [CrossRef]
9. Rajabi, M.; Nahavandchi, H.; Hoseini, M. Evaluation of CYGNSS Observations for Flood Detection and Mapping during Sistan and Baluchestan Torrential Rain in 2020. *Water* **2020**, *12*, 2047. [CrossRef]
10. Ruf, C.S.; Chew, C.; Lang, T.; Morris, M.G.; Nave, K.; Ridley, A.; Balasubramaniam, R. A New Paradigm in Earth Environmental Monitoring with the CYGNSS Small Satellite Constellation. *Sci. Rep.* **2018**, *8*, 8782. [CrossRef] [PubMed]
11. Tsyganskaya, V.; Martinis, S.; Marzahn, P. Flood Monitoring in Vegetated Areas Using Multitemporal Sentinel-1 Data: Impact of Time Series Features. *Water* **2019**, *11*, 1938. [CrossRef]
12. Tsyganskaya, V.; Martinis, S.; Marzahn, P.; Ludwig, R. Detection of Temporary Flooded Vegetation Using Sentinel-1 Time Series Data. *Remote Sens.* **2018**, *10*, 1286. [CrossRef]
13. Refice, A.; Zingaro, M.; D'Addabbo, A.; Chini, M. Integrating C- and L-Band SAR Imagery for Detailed Flood Monitoring of Remote Vegetated Areas. *Water* **2020**, *12*, 2745. [CrossRef]

Article

Application of Backpack-Mounted Mobile Mapping System and Rainfall–Runoff–Inundation Model for Flash Flood Analysis

Takahiro Sayama [1,*], Koji Matsumoto [2], Yuji Kuwano [3] and Kaoru Takara [4]

1   Disaster Prevention Research Institute, Kyoto University, Gokasho, Uji 611-0011, Japan
2   Graduate School of Engineering, Kyoto University, Nishikyo-ku, Kyoto 615-8540, Japan; matsumotokoji130513@gmail.com
3   Leica Geosystems K. K., 1-4-28, Mita, Minato-ku, Tokyo 108-0073, Japan; Yuji.kuwano@leica-geosystems.com
4   Graduate School of Advanced Integrated Studies in Human Survivability, Kyoto University, Yoshida-Nakaadachi 1, Sakyo-ku, Kyoto 606-8306, Japan; takara.kaoru.7v@kyoto-u.ac.jp
*   Correspondence: sayama.takahiro.3u@kyoto-u.ac.jp; Tel.: +81-774-38-4126

Received: 5 April 2019; Accepted: 4 May 2019; Published: 8 May 2019

Abstract: Satellite remote sensing has been used effectively to estimate flood inundation extents in large river basins. In the case of flash floods in mountainous catchments, however, it is difficult to use remote sensing information. To compensate for this situation, detailed rainfall–runoff and flood inundation models have been utilized. Regardless of the recent technological advances in simulations, there has been a significant lack of data for validating such models, particularly with respect to local flood inundation depths. To estimate flood inundation depths, this study proposes using a backpack-mounted mobile mapping system (MMS) for post-flood surveys. Our case study in Northern Kyushu Island, which was affected by devastating flash floods in July 2017, suggests that the MMS can be used to estimate the inundation depth with an accuracy of 0.14 m. Furthermore, the landform change due to deposition of sediments could be estimated by the MMS survey. By taking into consideration the change of topography, the rainfall–runoff–inundation (RRI) model could reasonably reproduce the flood inundation compared with the MMS measurements. Overall, this study demonstrates the effective application of the MMS and RRI model for flash flood analysis in mountainous river catchments.

Keywords: mobile mapping system; RRI model; high-water marks; inundation; Northern Kyushu floods; point clouds

1. Introduction

The spatial distribution of inundation depth represents the basic information required to understand the damage of flood disasters. For long-term and large-scale flooding, the use of satellite remote sensing is the most effective way to identify the extent of flood inundations [1–3]. Together with topographic data, some methods have been proposed to estimate the dynamics of flood inundation depths [4]. In the case of abrupt, small-scale flooding, such as flash floods in mountainous catchments, it has proven difficult to use satellite remote sensing techniques [5]. To compensate for this situation, detailed rainfall–runoff and flood inundation modeling has been performed to estimate the dynamics of flooding [5–7]. Such modeling is essential also for real-time flood forecasting and hazard mapping. Despite considerable advances in modeling, mainly due to the accurate and fine resolution of rainfall and topographic information in recent years, not much technological innovation has occurred in the investigation of high-water marks during post-flood field surveys. In fact, to carry out field measurements of high-water marks with a tape measure is laborious and time-consuming work [8]. Recently, a survey of high-water marks for research purposes was conducted using real-time kinematic GPS (RTK-GPS) [9]. The authors measured the levels of the high-water marks at multiple points for the Kinugawa flooding in the Kanto Region in September 2015 and estimated the spatial distribution of flood inundation depths by spatially interpolating the measurements and subtracting the ground elevation [10]. Since RTK-GPS measures the position of a point and its ground elevation using a single receiver, the measurements can be performed much more easily than in conventional detailed surveying methods. Nevertheless, it is similar to the conventional method in that it repeats the cycle of tasks of moving to the position where the high-water mark is visible, measuring the ground elevation with RTK-GPS, separately measuring the relative height from the ground level to the high-water mark and moving on to the next point. This takes a great deal of time and labor when targeting multiple points over a wide area.

To resolve this problem, the present study examines the application of a mobile mapping system (MMS) in high-water mark surveys. MMS is a surveying instrument consisting of a global navigation satellite system (GNSS), a rapid sequence camera, a laser scanner, and an inertial measurement unit (IMU), as a single package typically mounted on automobiles [11]. More recently, the system is carried in a backpack, which allows us to measure inaccessible sites or the interior of buildings. The MMS provides continuously photographed images and 3D point cloud data (if a laser scanner is embedded), from which we estimate the absolute position of an object, its height, relative length, etc. For application in post-disaster surveys, the Geographical Survey Institute (GSI) of Japan investigated the inundation height of the tsunami caused by the Great East Japan Earthquake using MMS [12]. For river management, the temporal change of riverine topography has been monitored with MMS mounted on a boat or a cart [13–18]. For post-flood surveying, however, there have been limited studies applying MMS in high-water mark measurement [9], and its feasibility and effectiveness are not clearly known. If it becomes possible to use MMS for surveys of high-water marks, it will not only enable efficient measurement of high-water marks at multiple points but also serve as a useful disaster record for hazard map creation, as well as verification of flood analysis in future, since this information can be stored as measurable continuous images.

This study applied a backpack-mounted MMS to the post-flood survey of the 2017 northern Kyushu floods. The measured results were compared with the records of direct measurement of high-water marks. In addition, we applied the RRI model [19–21] to the same region to simulate the dynamics of flood inundation with and without the effects of topographic change estimated by the MMS. The main objective of the present study is to examine the utility and feasibility of the MMS and RRI model to analyze information on flash flood disasters with the following specific objectives: (1) Investigate the applicability of a backpack-mounted MMS for post-flood surveys; (2) simulate the rainfall–runoff and flood inundation process at the catchment scale by the RRI model; (3) validate the simulation results with local information on the maximum flood inundation depths estimated by the MMS; and (4) discuss the impact of topographic change, also estimated by the MMS, on the flood inundation simulation.

2. Study Sites

In July 2017, a severe rain band formed over Northern Kyushu Island in Japan induced by a Baiu front and Typhoon No. 3. The rain band brought 829 mm of rainfall within 24 h between July 5, 8:00 and July 6 8:00, while the maximum hourly rainfall was 124 mm (on July 5 14:00–15:00) in Asakura City of Fukuoka Prefecture. The heavy rainfall caused many shallow landslides, which produced massive amounts of sediment and drift wood. The sediment deposits on the floodplains and storm water caused flooding along the tributaries of the Chikugo River (Figure 1). The disasters arising from the flood and sediment killed a total 37 people with 4 people declared missing in Fukuoka and Oita prefectures. Due to the floods, 288 houses were completely collapsed, 1095 and 44 houses were half- or partly-collapsed, respectively, and 172 and 1420 houses were flooded above or below floor levels, respectively.

Figure 2 shows the 24 h accumulated rainfall from midnight on July 5 to midnight on July 6. The rainfall is estimated from the synthetic rainfall products of C-band and X-band radars with 250 m resolution. The distribution of cumulative rainfall extends in an east–west direction, whose range corresponds with the shallow landslides indicated with orange color in Figure 1.

Among the tributaries of the Chikugo River Basin, we selected the Shirakitani River catchment (3.5 km^2) for intensive field measurement and simulation, as it was one of the most severely affected river catchments during the storm. The lower part of the Shirakitani River catchment is made up of plutonic granite, while its upper catchment and Western region belong to a metamorphic rock zone derived from mudstone. In the granite area, surface landslides of relatively shallow soil layers had occurred in several places, and as a result, large amounts of sediment and driftwood had buried rivers and riparian terraces (Figure 3).

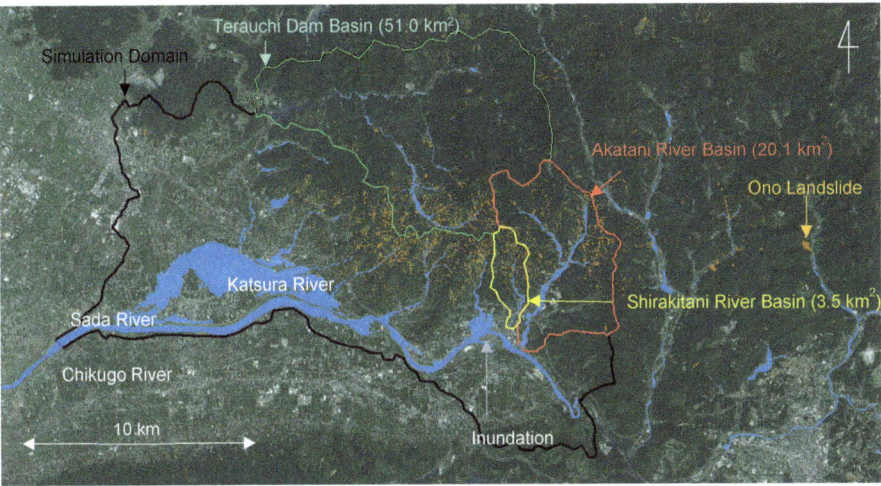

Figure 1. Disaster affected area of Northern Kyushu Island, including the Shirakitani River Basin.

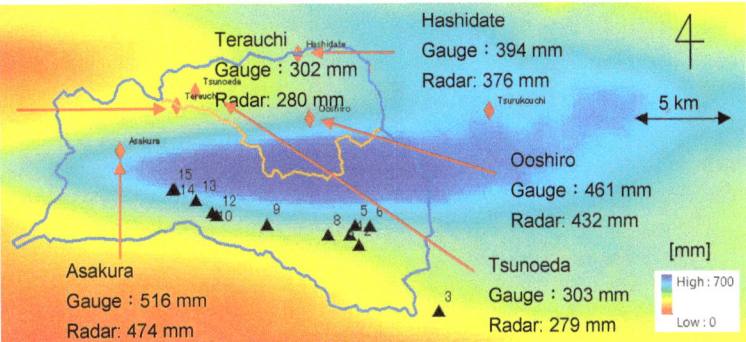

Figure 2. Cumulative rainfall distribution based on C- and X-band synthetic radar rainfall product (July 5 0:00 to July 6 0:00 in 2017). ▲ denotes the positions of our field survey for estimating the river geometry.

Figure 3. Situations of the flood disaster in the Shirakitani River Basin (as of March 25, 2018).

3. Methods

3.1. Mobile Mapping System (MMS)

In order to efficiently investigate the state of inundation and topographical changes, a field survey by MMS was carried out. The equipment used is the Leica Pegasus: Backpack (hereafter referred to as PB in this study) [22]. The PB is an MMS system that was conventionally installed in automobiles, now packed in a backpack, and comprises a GNSS, five rapid sequence cameras, two laser scanners, and an IMU (Figure 4). The greatest feature of the PB is its portability, and it is expected to be especially useful for information gathering and field surveys in disaster sites that cannot be accessed by automobiles. Since the weight of the system is about 12 kg, it can be carried as check-in baggage in aircraft and is thus convenient to use in disaster investigations.

The camera mounted on the PB has a charge coupled device (CCD) size of 2046 × 2046, a pixel size of 5.5 μm × 5.5 μm, and a maximum frame rate of 2 frames per second. With five cameras mounted on the sides and in the rear, continuous photographs can be taken at a rear angle of 200°. The laser scanner has a horizontal viewing angle of 270°, a vertical viewing angle of 30°, and a scanning speed of 300,000 points per second. Covering 360° horizontally with two laser scanners, continuous 3-dimensional (3D) data are created by synchronizing the image and 3D point cloud data of the same camera. Further, the installed GNSS supports GPS/Glonass/BeiDu/QZSS. The absolute positioning accuracy has a nominal value of 5 cm outdoors.

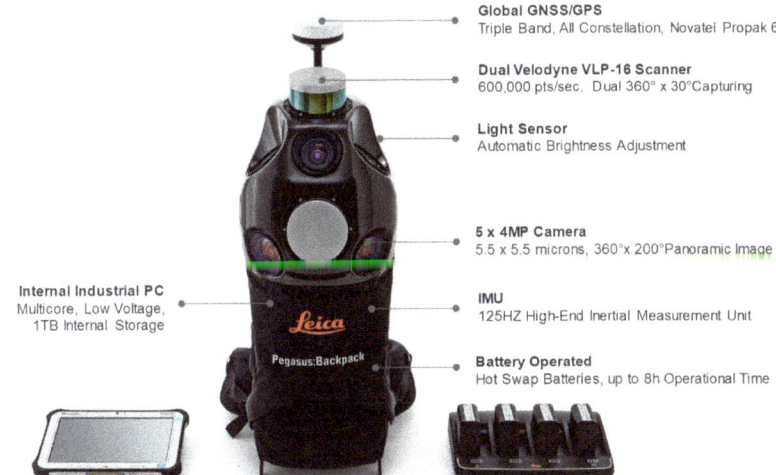

Figure 4. Leica Pegasus: External appearance of backpack (PB) (http://www.leica-geosystems.co.jp/jp/Leica-PegasusBackpack_106730.htm).

3.2. Field Survey with MMS

The field survey was conducted on September 12, 2017, and measurements were made by the author along with an engineer from Leica Geosystems (Figure 5). The survey route is indicated by the yellow line in Figure 6, and we made a round trip along the river, covering a distance of about 1.7 km. The time of the actual operation was approximately 3 h. Another survey using RTK-GPS was also conducted in parallel with the PB survey, which required more time than walking normally. In the survey using the PB, however, the investigation could proceed at the same speed as normal walking.

Figure 5. Field survey using PB (September 12, 2017: Photograph by Mr. Yoshiaki Ishida).

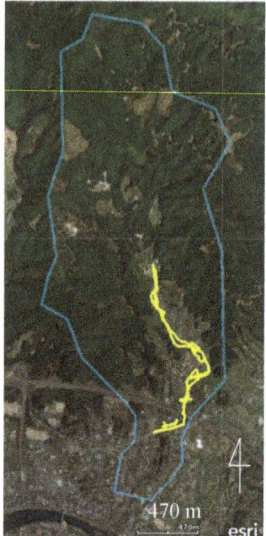

Figure 6. Survey route (yellow line) in the Shirakitani River Basin (blue line).

3.3. MMS Data Processing

The measured data were synchronized by image processing. The time taken for the image processing also depends on the amount of measurements and took several hours in this case. Once the processing was completed, the captured images can be continuously displayed, and by moving the cursor to a specific location on the screen, the absolute horizontal position and elevation of the object can be measured. In the present case, as we were able to identify high-water marks staining the side walls of buildings, we measured the absolute height of the high-water marks and the inundation depths at 12 points.

In the case of the northern Kyushu disaster, in addition to extracting the level of high-water marks, estimating the topographical change is also important for an understanding of the damage and to carry out the flood analysis, which will be described later. Since the PB creates 3D data for the region around the area covered by walking, analyzing this data will help to estimate the information in relation to the height and surface materials. Furthermore, a digital elevation model (DEM) with a spatial resolution of 10 m and without the influence of buildings and trees was created from the digital surface model (DSM). With the help of this model, it is possible to estimate the terrain after the disaster, and by comparing it with topographical information before the disaster, obtained by an aircraft laser profiler (LP), the depth of sediment deposition could be estimated (Table 1).

Table 1. Data processing for digital elevation models (DEMs).

DEMs	Procedures
Original DEM	Aircraft laser profiler (LP) produced a digital surface model (DSM). The digital elevation model (DEM) was then estimated by removing the effects of buildings and trees etc.
Post disaster DEM	Pegasus: Backpack (PB) produced DSM along the floodplain. DEM was then estimated by removing the effects of buildings and trees. The unmeasured area by the PB survey was filled by the above original DEM.

3.4. Rainfall–Runoff–Inundation (RRI) Model

This study applied the RRI Model to the Shirakitani River catchment to simulate flooding. The RRI Model is a 2-dimensional (2D) diffusive wave model that takes into account rainfall–runoff and river routing in the rivers and their exchange, representing overtopping flooding (see References [19–21] on details of the RRI model). The model treats slopes and river channels separately. For river grid cells, the model supposes that both slope and river are positioned. It applies the 2D diffusive wave model for slope water, while the 1D diffusive wave model is used for channel flow. For better representations of rainfall–runoff–inundation processes, the RRI model simulates also lateral subsurface flow, vertical infiltration flow, and surface flow. In the mountainous regions, the lateral subsurface flow is important. The model uses the discharge–hydraulic gradient relationship, which takes into account both saturated subsurface and surface flows. For flat terrain areas, we assume the vertical infiltration flow is essential and estimated by the Green–Ampt model. The flow interaction between the river channel and slope is estimated based on different overflowing formulae, depending on water-level and levee-height conditions. The model is applied not limited to floodplains but applicable to entire river basins.

The input to the model was the C- and X-band (CX) synthetic radar rainfall and the simulation period was from July 5 0:00 to July 7 0:00. The spatial resolution of the model was set at 10 m, and the model parameters were calibrated during the same period against the observed inflow to the Terauchi Dam (see Figure 1). In this model, the lateral flow and surface flow of the soil layer is reproduced on the mountainous forest slopes. We used two topographic datasets for the simulation, with one based on LP data by GSI, Japan, acquired prior to the disaster, and the other one based on the MMS survey created after the disaster. Although this model can reflect any cross-sectional shape, a simple rectangular cross section has been assumed here on account of the limited information. The width and depth were estimated by the empirical expressions $W = C_w A^{S_w}$ and $D = C_d A^{S_d}$, respectively. Here, A denotes the area of the catchment (km^2) at each location, and C_w, S_w, C_d, and S_d are parameters estimated from the river width and depth at the 8 sites measured in the downstream part of each branch, and set to 4.73, 0.58, 1.57, and 0.33, respectively.

4. Results and Discussions

4.1. Measurement of High-Water Marks by PB

This section presents the results of the field survey with the PB. Figure 7 shows a snapshot of the animation taken by the PB. High-water marks staining houses and fences are visible with adequate clarity in the image. Further, superimposing the color captured by the camera on the 3D point cloud data from the laser yields the images shown in Figures 8 and 9. Being 3D information, it is possible to verify the 3D features, even from angles that are not covered in the actual walking route (Figure 10). The method of displaying the camera image at the center and measuring the high-water mark and ground elevation of the point on the screen was found to be an efficient way to investigate high-water marks.

Figure 7. Measurement data by PB (**Left**: Laser point cloud data colored with reflected intensity, **Right**: Stereo image).

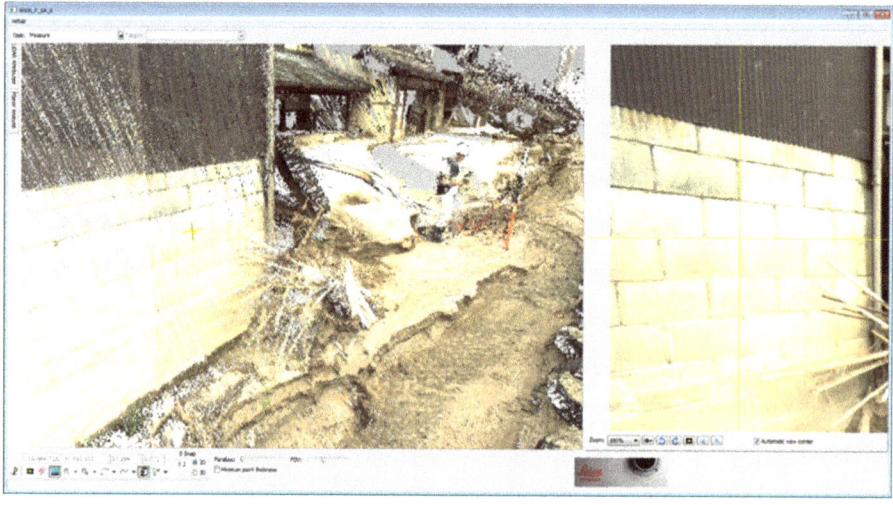

Figure 8. Flood traces on sidewall of building.

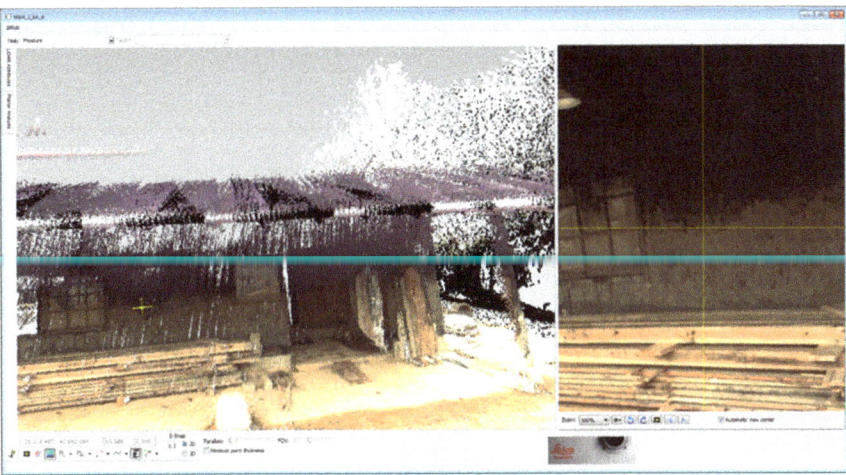

Figure 9. Results of superimposing (**left**) the color of the image captured by the camera (**right**) on the laser point cloud.

Figure 10. Bird's-eye view of the 3-dimensional image created from PB laser point cloud.

Table 2 shows the location and inundation depth of high-water marks estimated from the images generated with the PB. The inundation depth estimated from the PB lies in the range of 0.34 m to 1.15 m. It must be noted that this inundation depth was measured from the ground height (G. Level) after the flooding and the raising of the ground level by sediment deposition is not included in this value.

In order to confirm the accuracy of the inundation depth measured using the image, a field survey was conducted once again on March 25, 2018. As shown in Table 2 (Obs: W. Depth), the difference between the PB and field survey was less than 0.11 m, except in the case of two locations, which will be discussed later. As the high-water marks also consisted of water mixed with sediments, there was some uncertainty about which points should be regarded as high-water marks. However, the relative height measurement by the PB was found to have sufficient accuracy with a mean error (ME) of −0.08 and root mean square error (RMSE) of 0.14 m.

Table 2. Comparison of ground elevation (Z), inundation depths (wl), and field survey (obs) after the disaster, as measured by PB.

ID	Position		PB			LP	Obs.	RRI	Evaluation		
Item	Position		G. Level	W. Level	W. Depth	G. Level	W. Depth	W. Depth	G. Level	W. Depth	W. Depth
Notation	LON	LAT	Z_pb	wl_pb	hs_pb	Z_lp	hs_obs	hs_sim	Z_pb-Z_lp	hs_pb-hs_obs	hs_sim-hs_obs
1	130.82061	33.36678	55.97	57.12	1.15	56.02	1.20	1.16	−0.05	−0.05	−0.04
2	130.82091	33.36682	55.85	56.62	0.77	56.60	0.87	0.91	−0.75	−0.10	0.04
3	130.82266	33.36851	64.33	64.91	0.58	65.11	0.60	0.18	−0.78	−0.02	−0.42
4	130.82246	33.36854	64.30	65.01	0.71	64.51	0.80	0.37	−0.21	−0.09	−0.43
5	130.82317	33.37125	76.87	77.40	0.53	76.87	0.58	0.36	0.00	−0.05	−0.22
6	130.81992	33.37503	106.81	107.15	0.34	106.71	0.28	0.15	0.10	0.06	−0.13
7	130.81977	33.37560	110.49	111.05	0.56	110.07	0.55	0.17	0.42	0.01	−0.38
8	130.81971	33.37569	110.81	111.58	0.77	110.44	0.88	0.43	0.37	−0.11	−0.45
9	130.82261	33.36957	68.06	68.76	0.70	67.75	0.95	0.39	0.31	−0.25	−0.56
10	130.82075	33.36681	56.16	57.07	0.91	56.17	1.28	1.09	−0.01	−0.37	−0.19
11	130.82102	33.36688	57.01	57.43	0.42	57.07	0.50	0.7	−0.06	−0.08	0.20
12	130.82271	33.36963	68.40	68.91	0.51	68.53	0.48	0.36	−0.13	0.03	−0.12
								ME	−0.07	−0.08	−0.23
								RMSE	0.37	0.14	0.31

Ground height: G. Level (Z_pb: Estimation by PB, Z_lp: Estimation by LP elevation data). Inundation level: W. Level (wl_pb: Estimation by PB, hs_obs: Field measurement, hs_sim: Estimation by RRI model). LON: longitude and LAT: latitude.

The PB values of measurement numbers 9 and 10 are, respectively, 0.25 m and 0.37 m less than the field survey values. In measurement 9, the selection of the measurement site posed a problem, and the splashing of large amounts of deposits mentioned earlier could be responsible for the overestimation of the field measurement. On the other hand, in the case of measurement 10, some of the sediment that accumulated in front of the house had been removed by the time of the survey conducted on March 25, 2018. Since the field measurement measured the depth at a location from where sediment had been removed up to the level of the high-water marks, it is likely to be greater than the inundation depth estimated from the PB image.

Since reference points, such as benchmarks, had not been used in conducting the present survey, it is difficult to discuss, in strict terms, the accuracy of the absolute height estimated by the PB. However, the comparison with the results of the 10 points measured separately using RTK-GPS showed the RMSE in the horizontal direction to be 0.12 m and the RMSE in the vertical direction to be 0.11 m. The accuracy of the RTK-GPS equipment was approximately 0.015 m in the horizontal direction and 0.02 m in the vertical direction. Hence, the PB estimation may be regarded as being sufficiently accurate, considering that it is obtained by simply specifying the target on the screen.

4.2. Estimation of Topographic Changes by PB

A DEM with 10 m spatial resolution was prepared based on the 3D point cloud data measured with the PB. The DEM is limited to a range of about 150 m laterally on either side of the centerline. To create a wide area DEM, it is necessary to carry out longitudinal and lateral surveys. Nevertheless, in creating a DEM of the floodplain along small and medium rivers as in the present case, the range that is generally necessary for exploring along the river was covered.

Figure 11 shows the difference between the topography after the disaster estimated by the PB and by the LP data before the disaster. In much of the measurement range of the PB, the height is observed to rise due to sediment deposition, reaching up to 4 to 5 m in some places in the range of approximately 100 m from the river, where the amount of deposition is particularly high. This is consistent with the scene shown in the photograph in Figure 3 (lower left), where a portion of the house up to the first floor is completely buried under sediments.

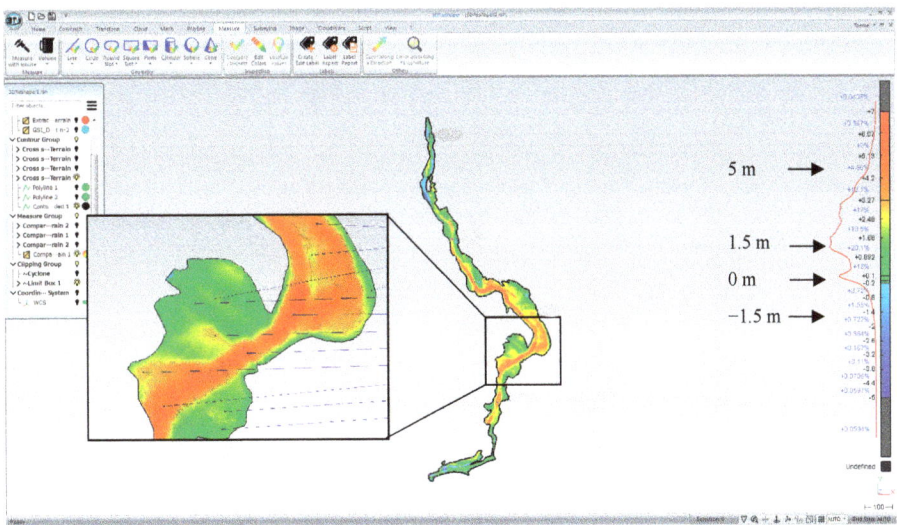

Figure 11. Topographical change estimated by PB and DEM before the disaster (positive values denote deposition and negative values denote erosion).

4.3. Verification of Rainfall–Runoff–Inundation Simulation with the PB Measurement

We conducted an integrated analysis of rainfall, runoff, and inundation in the Shirakitani River catchment. The inundation depth measurement at multiple points by the PB is useful information for the verification of such flood analysis. Further, the current disaster has resulted in topographical changes due to the massive flow of sediments, and we also discuss here the importance of considering its impact when carrying out flood analysis.

CX synthetic radar rainfall was input to the RRI model to estimate the runoff of the small rivers. The parameters of the model were calibrated with the observed values of the inflow at the Terauchi Dam (Figure 12a). For the present case, the observed dam inflow was well reproduced by setting a relatively thin layer of soil depth (0.6 m) without percolation from the soil layer to the underlying bedrock. Figure 12b shows the estimated hydrograph at the downstream point of Shirakitani River catchment.

Figures 13b and 14b show the simulated maximum inundation depths in the downstream part and whole catchment, respectively, as estimated by the RRI model. For comparison, we also present the flood inundation extent estimated by GSI, Japan, which determined the maximum flood extent based on aerial photos and field investigation. The comparison suggests that the RRI model underestimates the extent estimated by GSI. This is mainly due to ignorance of the topographic change caused by the sedimentation during the storm event. The maximum water depth is estimated to be about 1.5 m, which only explains the filling of the downstream river channel and does not explain flooding that could fill the floor of the valley, which had actually occurred. This suggests that impediments to the downward flow of the river due to the accumulation of sediments or topographical changes in the flood plains had an impact on the flooding.

As mentioned earlier, the PB survey provides information about the 3D topography of the surroundings, and hence, the DEM after the disaster can be estimated from the result. Figure 13a shows the amount of topographical change estimated by subtracting the predisaster DEM by the GSI from the DEM estimated by the PB after the disaster. This result is the same as in Figure 11 above but has been obtained by resampling the data with the resolution of 10 m corresponding to the calculation grid of the RRI model. As already mentioned, the analysis result using the DEM before the disaster does not include the inundation of the surroundings (Figures 13b and 14b). The result of the analysis using the topography after the disaster corresponds well with the inundation range given by the GSI, as can be seen in Figures 13c and 14c (blue lines in the figures).

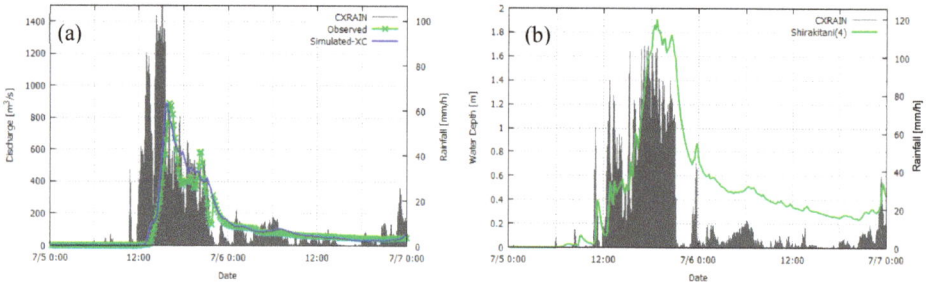

Figure 12. (a) RRI model calibration with observed dam inflow at Terauchi Dam, (b) simulated water depths at the downstream of Shirakitani River.

In the graph in Figure 15, the direct measurement of the maximum inundation depth at the 12 points mentioned above have been plotted on the horizontal axis, and the estimation from the PB and RRI models is plotted on the vertical axis. Comparing the field survey and PB values, the mean error (ME) was found to be −0.08 m, and the RMSE was 0.14 m. On the other hand, the comparison between the field survey and the RRI model shows that the ME is −0.23 m and the RMSE is 0.31 m, indicating that the model tends to underestimate the maximum inundation depth. In the analysis, only the rainfall, runoff, and inundation have been calculated based on the topographical changes after the flooding, whereas the outflow and sedimentation contain a mixture of water and sand deposits, which could be the cause of the underestimation.

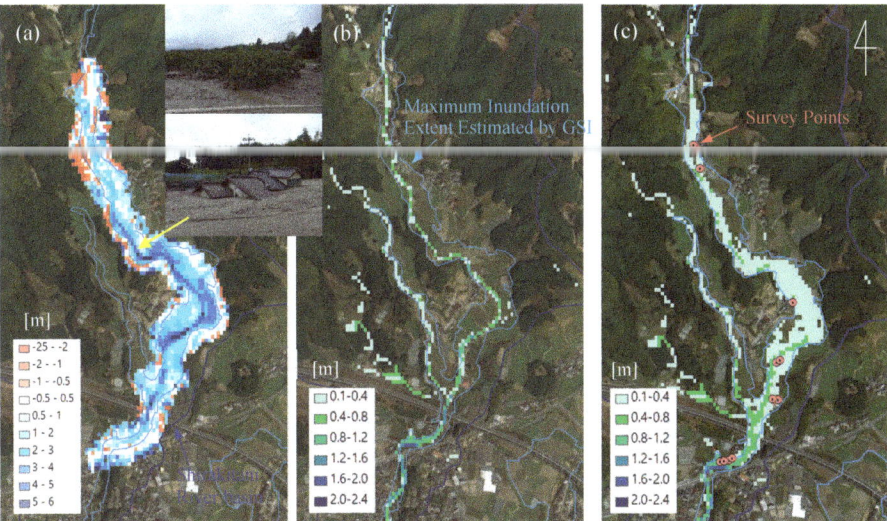

Figure 13. In the downstream part of the Shirakitani River basin: (**a**) Topographical change (Difference between the DEM by mobile mapping system (MMS) after the disaster and DEM by the Geographical Survey Institute before the disaster), maximum inundation depth distribution estimated by the RRI model (**b**) with the original DEM and (**c**) with the new DEM created by the PB analysis.

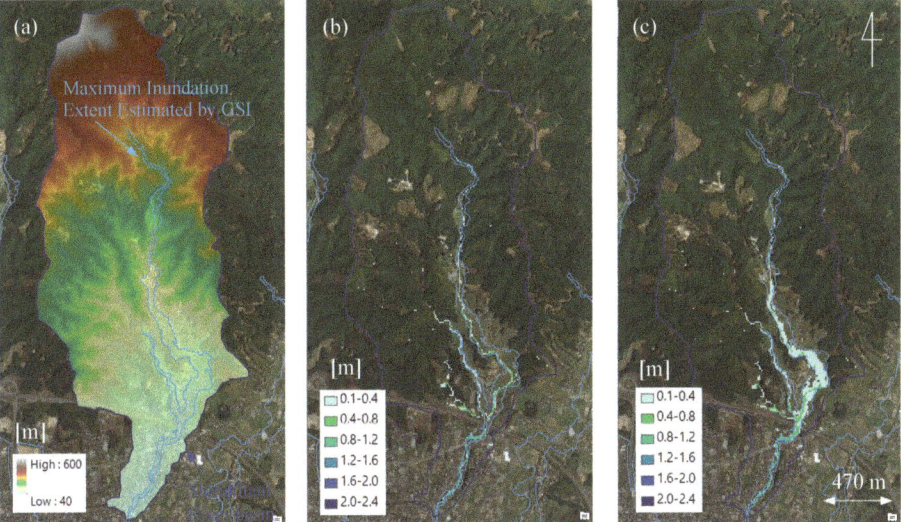

Figure 14. Same as the Figure 13, but with (**a**) original DEM and (**b**), (**c**) covering for the entire river basin.

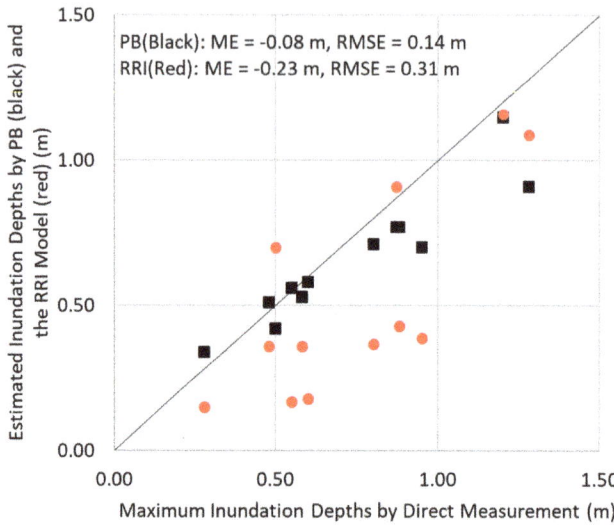

Figure 15. Comparison of maximum inundation depths estimated by PB (black solid squares) and the rainfall–runoff–inundation (RRI) model (red circles) against the direct measurements.

4.4. On the Potential Use and Limitations of MMS and the Modeling for Post Flood Surveys

Based on the above results, we describe the future prospects for the use of MMS in field surveys immediately after a disaster. In the survey by MMS, since it is possible to concentrate only on collection of images and 3D point cloud data while working at the site, large areas can be investigated quickly compared to the conventional high-water mark surveys that are conducted by stopping at specific places. Moreover, the work is divided into data collection in the field and measurement work at the office, which is a great advantage, especially in surveying disaster areas. Furthermore, since data from MMS can be archived, in addition to becoming a record of the damage, it is particularly useful in verification of the flood analysis as the position and height of any desired place can be measured at any time.

At the start of this research, we had assumed the use of vehicle mounted MMS, but we failed to do so because the disaster-affected area was inaccessible by vehicles. Such vehicle-mounted MMS may be used in wide plain areas. For the disaster affected site here, the backpack type equipment was the most useful alternative. Not all MMSs are equipped with a laser scanning system, and instead, they estimate a relative distance by image processing, which can limit the data size. In the case of having to determine in detail the absolute height of high-water marks, the MMS with a laser scan, such as the PB, is more effective.

Further case studies of the presented approach are definitely required. As described above, the flash flooding in Northern Kyushu Island in July 2017 is characterized with the massive landslides and sedimentation along the floodplain. With high sediment concentration of the inundated water, the visibilities of flood marks may be higher than other cases without sediment concentration. In addition, such a destructive event prevented local residents from immediate recovering activities including clearing flood marks from their houses. These conditions are not necessarily satisfied in other flood events, which may require even rapid post flood surveys.

On the other hand, the sedimentation with the significant topographic changes made it difficult for the flash flood simulation. Due to the current model capability as well as our study scope, we demonstrated the RRI model application with topographic information before and after the disaster event. We suppose this is the first case study using MMS for estimating topographic changes after flash floods in a mountainous region. The MMS survey was inevitable for estimating the topographic change, which was necessary also for the flash flood simulations. Nevertheless, for the better prediction of such disasters, it is important to simulate rainfall–runoff, flood inundation, and sedimentation transportation in a more integrated manner.

5. Conclusions

In this study, we examined the utility and feasibility of using the backpack type MMS (Pegasus: Backpack, PB of Leica Geosystems). We made a 1.7 km long round-trip survey along the Shirakitani River while carrying the equipment and collected continuous images and a 3D point cloud data. Our conclusions are summarized below.

(1) High-water marks due to the flood disaster staining building walls could be clearly identified from the continuous images and 3D point cloud data of the PB. If the level of a high-water mark is specified on the screen, its position and water depth can be measured.

(2) As no special operations are required, except for working with a tablet terminal to make color adjustments to captured images, the survey could proceed at the same speed as walking.

(3) The RMSE of maximum inundation depths measured by the PB was 0.14 m according to the comparison with direct measurements at twelve points. Considering the fact that high-water marks staining the buildings are affected by splashing, the error is thought to be within the permissible range. We believe that the PB has sufficient accuracy to carry out high-water mark measurements.

(4) A new DEM was also created using the PB measurements. It identified a maximum 4 to 5 m of sediment deposits in some places along the river. The information on topographic change was essential for our RRI simulation. The simulation, using the post-disaster topographic data, was found to be closer to the actual flooded area.

(5) The results of the flood analysis by the RRI model were compared with the field survey of inundation depths. The comparison with the field survey showed an ME of −0.23 m and RMSE of 0.31 m, indicating that the model tended to slightly underestimate the maximum inundation depth. In the analysis, only the rainfall runoff and inundation have been calculated based on the topographical changes after the flooding, whereas in reality, the outflow and sedimentation contain a mixture of water and sand deposits, which may have affected the results of the analysis.

Overall, this study demonstrated the high potential of the wearable MMS and the RRI model in post-flood analysis. Survey-related technologies, such as MMS and drones, have been making impressive advances, and it is important to gather and archive information about disaster-affected areas utilizing such state-of-the-art technology.

Author Contributions: Conceptualization, T.S.; methodology, T.S. and Y.K.; software, Y.K.; validation, T.S., K.M.; formal analysis, T.S.; investigation, T.S.; resources, K.T.; data curation, K.M.; writing—original draft preparation, T.S.; writing—review and editing, T.S.; visualization, T.S., K.M. and Y.K.; supervision, K.T.; project administration, T.S.; funding acquisition, T.S. and K.T.

Funding: This study was supported by the River Fund of The River Foundation, Japan, JSPS KAKENHI Grant Number 15H04051 (PI: T. Sayama) and JSPS Grant-in-Aid for Special Purposes 17K20140 (PI: J. Akiyama).

Acknowledgments: We acknowledge the technical support by Leica Geosystems K. K., Japan for the technical support of Pegasus: Backpack.

Conflicts of Interest: The authors declare no conflict of interest.

References

1. Schumann, G.; Bates, P.D.; Horritt, M.S.; Matgen, P.; Pappenberger, F. Progress in integration of remote sensing–derived flood extent and stage data and hydraulic models. *Rev. Geophys.* **2009**, *47*. [CrossRef]
2. Bates, P.D. Integrating remote sensing data with flood inundation models: How far have we got? *Hydrol. Process.* **2012**, *26*, 2515–2521. [CrossRef]
3. Rahman, M.S.; Di, L. The state of the art of spaceborne remote sensing in flood management. *Nat. Hazards* **2016**, *85*, 1223–1248. [CrossRef]

4. Nguyen, N.Y.; Ichikawa, Y.; Ishidaira, H. Estimation of inundation depth using flood extent information and hydrodynamic simulations. *Hydrol. Res. Lett.* **2016**, *10*, 39–44. [CrossRef]
5. Elfeki, A.; Masoud, M.; Niyazi, B. Integrated rainfall–runoff and flood inundation modeling for flash flood risk assessment under data scarcity in arid regions: Wadi Fatimah basin case study, Saudi Arabia. *Nat. Hazards* **2016**, *85*, 87–109. [CrossRef]
6. Kvočka, D.A.; Ahmadian, R.; Falconer, R. Flood Inundation Modelling of Flash Floods in Steep River Basins and Catchments. *Water* **2017**, *9*, 705. [CrossRef]
7. Mashaly, J.; Ghoneim, E. Flash Flood Hazard Using Optical, Radar, and Stereo-Pair Derived DEM: Eastern Desert, Egypt. *Remote Sens.* **2018**, *10*, 1204. [CrossRef]
8. Koenig, T.A.; Bruce, J.L.; O'Connor, J.E.; McGee, B.D.; Holmes, R.R., Jr.; Hollins, R.; Forbes, B.T.; Kohn, M.S.; Schellekens, M.F.; Martin, Z.W.; et al. *Identifying and Preserving High-Water Mark Data*; U.S. Geological Survey: Reston, VA, USA, 2016; 47p.
9. Meesuk, V.; Vojinovic, Z.; Mynett, A.E. Extracting inundation patterns from flood watermarks with remote sensing SfM technique to enhance urban flood simulation: The case of Ayutthaya, Thailand. *Comput. Environ. Urban Syst.* **2017**, *64*, 239–253. [CrossRef]
10. Sayama, T.; Takara, K. Estimation of Inundation Depth Distribution for the Kinu River Flooding by 2015.09 Kanto-Tohoku Heavy Rainfall. *J. Jpn. Soc. Civ. Eng. Ser. B1 (Hydraul. Eng.)* **2016**, *72*, I_1171–I_1176. [CrossRef]
11. Lauterbach, H.A.; Borrmann, D.; Hess, R.; Eck, D.; Schilling, K.; Nuchter, A. Evaluation of a Backpack-Mounted 3D Mobile Scanning System. *Remote Sens.* **2015**, *7*, 13753–13781. [CrossRef]
12. Koarai, M.; Okatani, T.; Nakano, T.; Nakamura, T.; Hasegawa, M. Geographical Information Analysis of Tsunami Flooded Area by the Great East Japan Earthquake Using Mobile Mapping System. *Int. Arch. Photogramm.* **2012**, *39-B8*, 27–32. [CrossRef]
13. Vaaja, M.; Hyyppa, J.; Kukko, A.; Kaartinen, H.; Hyyppa, H.; Alho, P. Mapping Topography Changes and Elevation Accuracies Using a Mobile Laser Scanner. *Remote Sens.* **2011**, *3*, 587–600. [CrossRef]
14. Saarinen, N.; Vastaranta, M.; Vaaja, M.; Lotsari, E.; Jaakkola, A.; Kukko, A.; Kaartinen, H.; Holopainen, M.; Hyyppa, H.; Alho, P. Area-Based Approach for Mapping and Monitoring Riverine Vegetation Using Mobile Laser Scanning. *Remote Sens.* **2013**, *5*, 5285–5303. [CrossRef]
15. Williams, R.D.; Brasington, J.; Vericat, D.; Hicks, D.M. Hyperscale terrain modelling of braided rivers: fusing mobile terrestrial laser scanning and optical bathymetric mapping. *Earth Surf. Proc. Landf.* **2014**, *39*, 167–183. [CrossRef]
16. Leyland, J.; Hackney, C.R.; Darby, S.E.; Parsons, D.R.; Best, J.L.; Nicholas, A.P.; Aalto, R.; Lague, D. Extreme flood-driven fluvial bank erosion and sediment loads: Direct process measurements using integrated Mobile Laser Scanning (MLS) and hydro-acoustic techniques. *Earth Surf. Proc. Landf.* **2017**, *42*, 334–346. [CrossRef]
17. Entwistle, N.; Heritage, G.; Milan, D. Recent remote sensing applications for hydro and morphodynamic monitoring and modelling. *Earth Surf. Proc. Landf.* **2018**, *43*, 2283–2291. [CrossRef]
18. Lotsari, E.S.; Calle, M.; Benito, G.; Kukko, A.; Kaartinen, H.; Hyyppä, J.; Hyyppä, H.; Alho, P. Topographical change caused by moderate and small floods in a gravel bed ephemeral river—A depth-averaged morphodynamic simulation approach. *Earth Surf. Dyn.* **2018**, *6*, 163–185. [CrossRef]
19. Sayama, T.; Ozawa, G.; Kawakami, T.; Nabesaka, S.; Fukami, K. Rainfall–runoff–inundation analysis of the 2010 Pakistan flood in the Kabul River basin. *Hydrol. Sci. J.* **2012**, *57*, 298–312. [CrossRef]
20. Sayama, T.; Tatebe, Y.; Tanaka, S. An emergency response-type rainfall-runoff-inundation simulation for 2011 Thailand floods. *J. Flood Risk Manag.* **2015**, *10*, 65–78. [CrossRef]
21. Sayama, T.; Tatebe, Y.; Iwami, Y.; Tanaka, S. Hydrologic sensitivity of flood runoff and inundation: 2011 Thailand floods in the Chao Phraya River basin. *Nat. Hazards Earth Syst. Sci.* **2015**, *15*, 1617–1630. [CrossRef]
22. Leica. Leica Pegasus: Backpack Wearable Mobile Mapping Solution. Available online: https://leica-geosystems.com/products/mobile-sensor-platforms/capture-platforms/leica-pegasus-backpack (accessed on 21 September 2018).

© 2019 by the authors. Licensee MDPI, Basel, Switzerland. This article is an open access article distributed under the terms and conditions of the Creative Commons Attribution (CC BY) license (http://creativecommons.org/licenses/by/4.0/).

Article

Geometry-Based Assessment of Levee Stability and Overtopping Using Airborne LiDAR Altimetry: A Case Study in the Pearl River Delta, Southern China

Xianwei Wang [1,2,3,*], Lingzhi Wang [1] and Tianqiao Zhang [4]

1. School of Geography and Planning, Sun Yat-sen University, Guangzhou 510275, China; wanglzh6@mail2.sysu.edu.cn
2. Guangdong Provincial Engineering Research Center for Public Security and Disasters, Guangzhou 510275, China
3. Southern Marine Science and Engineering Guangdong Laboratory (Zhuhai), Zhuhai 519080, China
4. Guangzhou Jiantong Surveying, Mapping and Geoinformation Technology Co. LTD., Guangzhou 510520, China; 136226994@qq.com
* Correspondence: wangxw8@mail.sysu.edu.cn; Tel.: +86-20-84114623

Received: 18 December 2019; Accepted: 28 January 2020; Published: 2 February 2020

Abstract: Levees are normally the last barrier for defending flood water and storm surges in low-lying coastal cities. Levees in a large delta plain were usually constructed in different time and criteria and have been changing with age as well. Fast and quantitative assessment of levee stability is critical but faces many challenges. This study designs a scoring approach to quickly assess levee stability and overtopping threats with geometric parameters from airborne Light Detection and Ranging (LiDAR). An automated procedure is developed to extract levees geometric parameters from 0.5 m grid LiDAR elevation, such as crown height, width and landside slope. The surveyed levee is seated in the Hengmen waterway in the Pearl River Delta, Southern China. Results show that the stability index using the assessment scores is higher than and superior to the common qualified rates adopted in previous studies. The qualified rate is defined as the count percentage that each parameter meets the designed criteria, while the assessment score proposed in this study assigns different credits to those below/above the designed criteria. The continuous crown heights provide detailed information on levee overtopping threats. The crown heights of levee A and B are above the designed elevation and the flood stage (4.5 m) in a 200-year return period. The crown heights of levee C, D and E are generally lower than 4.5 m and vary in a large range on different sections. The middle section of levee E for the harbor and dock area has front elevation slightly below the flood stage (3.54 m) in a 20-year return period. Moreover, the high precision LiDAR altimetry data reveal various morphological modifications in all levees, such as natural subsidence and artificial modifications, which greatly reduce levees safety and are severe threats to the community. The procedures and assessment approach developed in this study can be easily applied for levees fast assessment in the entire Pearl River Delta and somewhere else, thus offer a suitable mitigation suggestion ahead of levee failure or overtopping.

Keywords: LiDAR; geometric parameters; levee stability; overtopping; Pearl River Delta

1. Introduction

Levees play a crucial role in defending flood water in the low-lying coast. Coastal cities often face flood threats caused by river fresh flood and ocean storm surges. The rising sea level under the context of global warming aggravates the flooding risk of coastal cities, such as the low-lying Pearl River Delta of Southern China [1]. Hydraulic engineering measures such as sea walls and levees can effectively resist flood shocks, while they are usually the last barrier to protect the lives and properties of local residents [2–4]. Levees cannot completely exclude flood disasters. Living behind a levee faces unique flood risks since levees are designed to reduce the impact of a flood event at certain scale [5]. However, the floodplain communities often underestimate the flood risk by the false concept that flood risk has been eliminated by the levee [6]. Moreover, frequent small floods can cause erosion, submergence, sedimentation of levees, finally resulting in levee failure or overtopping as a large flood event occurs, such as the catastrophic event of New Orleans hit by the Hurricane Katrina in 2005 [7]. About one-third of floods were related to levee breaks in the United States [8].

There are complex levee systems with several thousand kilometers in the low-lying Pearl River Delta in the southern China. Levees had been constructed in different time and criteria based on the flood-control level planned in their protecting area. The oldest levees were first constructed from natural dikes over 1000 years ago and have been stacked up in different periods. The design standards have been improved to protect the fast-developing towns, commercial and industry areas especially during the past four decades. Meanwhile, the working conditions of levees always turn worse with age due to natural subsidence, river scoring and human activities, resulting in morphological modification and degradation of the flood defense capability or higher flood risk [7]. Assessment of levee stability is a pressing and laborious task for the local levee management. Most levee assessments were visual and qualitative check by ground cruise prior to and during the flood season in the Pearl River Delta.

Regular assessment of levee stability and flood defense capacity is critical to guarantee the community safety behind levees. Levee assessment of stability and overtopping requires geometric characteristics, such as crown elevation, width, and slopes on the waterside and landside [7,9]. These parameters are traditionally obtained by ground physical surveying across a levee transect, which is time-consuming and laborious but still widely adopted in many places. The high resolution satellite images such as QuickBird and IKONOS demonstrate their potential in detecting levee slides by visual inspection and slide detection algorithms including image classification and spatial modeling along the Mississippi River in Bolivar County, Mississippi, US [10]. Neuner [11] identified two levee slide areas by visual inspection of the spectrally enhanced imagery. The high resolution spaceborne and airborne multispectral images are often applied to monitor the levee vegetation cover and soil water content, and the high moisture contents on the levee inclined surface and toes shows a close association with levee slides [12]. Vegetation indices, such as Normalized Difference Vegetation Index (NDVI), Red edge Vegetation Stress Index (RVSI) and Red Edge Position Index (REP), are developed from the airborne hyperspectral imagery to predict shallow surficial failures in the Mississippi River since levee slide-affected areas are often characterized by anomalous vegetation [13]. The high-resolution elevation data retrieved from airborne Light Detection and Ranging (LiDAR) enable fast and large-scale examination of the levee physical conditions [14]. Several approaches have been developed to extract the geometric parameters of levees from airborne LiDAR elevation data, such as the least-cost path (LCP), Flip 7 software [14], slope classification, morphological filtering, cluster algorithm and break line detection [15]. Although several algorithms had been developed, it is still not trivial to develop a completely automated approach to extract these parameters from LiDAR data [16].

The common method for levee assessment is to compare the geometric parameters against the design criteria. Casas et al. [7] developed a levee stability index by comparing the current geometric parameters at each levee transect with the design standards in the Sacramento–San Joaquin River Delta, California, US. Those meeting the minimum levee geometric criteria are labeled as in good conditions, and others are in poor condition, in terms of levee height, crown width, waterside slope, and landside slope according to the Geotechnical Levee Practice standards [17]. Adding together, levee transects are classified as very good (all 4 hits), good (3 hits), fair (2 hits), poor (1 hit) and very poor (all fail). Choung [9] assess the risk of levee overtopping by comparing the levee height against the designed flood level plus a 2 m freeboard in Nakdong River Basins, South Korea. Those segments lower than the designed flood level (1:100 year) are flagged as an area with a risk of overtopping. Palaseanu-Lovejoy et al. [16] evaluate the levee crest elevation with the federal levee standards, the 2010 refined flow line elevation above the 1:100-year flood stage plus a freeboard of 0.91 m. The error (0.24 m) of DEM data at the 95% confidence level is also considered in the comparison.

All the aforementioned cases compare the levee geometric parameters with only one standard value partially because of their small study area. However, levees in a large delta plain were usually constructed in different criteria based on their protecting targets, which undergo fast changes, such as in the Pearl River Delta. Levee degradation is usually accompanied by the morphometric modification from its initial design standards, such as levee crest subsidence, crown narrowing and slope steepen. Those morphometric modifications will eventually affect the levee integrity and even result in levee failure [7]. The one criterion approach could not meet the need of levee assessment in a large and complex levee system. More information is needed in assessing the actual performance of each levee segments beyond the binary assessment result, "good/hit" or "poor/fail", especially for those segments below the designed standard.

In the context of global warming and new national development strategy in the great bay area of Guangdong-Hong Kong- Macao, levees will bear on more and more important roles in defending the flood water and storm surges. Fast and quantitative assessment of levee stability and flood risk is in great demand as the levee ages. The primary objective of this study is to develop a procedure to automatically extract levees geometric parameters from airborne LiDAR data, and then to design a scoring approach to assess their flood defense capacity, i.e., levee stability and overtopping risk, according to levees geometric parameters, the construction code, the sea level rise rate, and the designed water levels at several flood frequencies. Both can be used together to quickly assess the long and complex levee systems in the Pearl River Delta, Southern China and other regions in the world.

2. Study Levees and Data Processing

2.1. Study Levees

The levees surveyed in this study are along the Hengmen waterway in the Pearl River Delta, Southern China at 22°34′–22°35′ N and 113°24′–113°38′ E (Figure 1). The Hengmen is one of the eight outlets for the Pearl River estuary. It discharges flood water to the eastern bay of Lingdingyang mostly from the Xijiang via the Jiya and Xiaolan waterways in the west. The surveyed levees have total length of 44.5 km and are divided into five segments from A to E based on their natural features, protecting targets and designed criteria. All the five segments are particularly selected to represent different types of the levee system. The segments A and B are standard large levees with concrete levee crown and waterside surface. The segment A is the southern part of the Minzhong-Sanjiao (Minsan) joint levee (Grade III), which was initially built in 1958 and had been finished rebuilding in 2008. It protects the towns of Minzhong and Sanjiao with total population of over 250,000 and an area about 190 km^2 in the city of Zhongshan, Guangdong Province, China. The segment B is a small eastern section of the Zhongshan-Shunde (Zhongshun) joint levee (Grade II), which was first enclosed in 1970s and rebuilt in 1990–1992 to defend a 1:50-year flood event. It protects over 640,000 people and areas of 700 km^2. The design standards of levees in segments C and D are being updated from defending the 1:20-year to 1:50-year flood event to protect the lately planned business districts of Huoju and Cuiheng in the city of Zhongshan. The segment E is a 7-km long dock and commercialized area without standard levee. Only surface height and the waterside slopes can be obtained in assessment. The further south of the segments C and E are the uplift hills of Wuguishan, which controls the sediment plain of this area in Zhongshan. The low-lying area protected by the segments C and E is much smaller than those of A and B, but is a fast-developing business district during past two decades. The entire Cuiheng district was reclaimed from the tidal beach and protected by the levee D, whose previous designed criteria was quite low and are being improved to resist a 1:50-year storm surge and tidal level. Only one third of the Cuiheng levee was surveyed in this study.

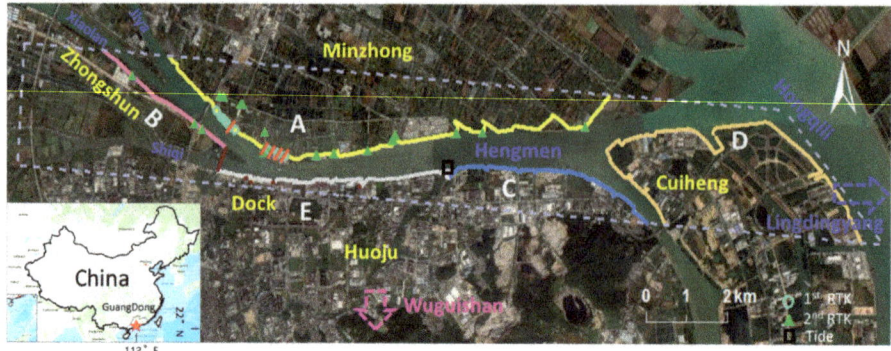

Figure 1. The airborne LiDAR surveyed areas (gray dashed line) in the Hengmen waterway of the Pearl River Delta, Zhongshan, Guangdong Province, China. The triangles, circles and red lines are the locations of in situ elevation measurement using GPS-RTK. The black square is the site of the Hengmen tidal station. The surveyed levees are divided into five segments from A to E.

2.2. Data Collection and Processing

The LiDAR data were collected around 13:00 during the low tide period on 9 December 2016. It was selected for a particular low water level for better surveying the levee toe and the tidal beach. The data were acquired from the airborne Harrier 68i LiDAR system from the Trimble Company, USA and a digital camera, the Rollei Metric AIC Pro (60 million pixels) from TopoSys Company in Germany. The aircraft is the Bell-206 helicopter. The flight height is 400 m above ground. The image pixel size is 0.05 m, and the laser point cloud density is 16 points/m^2 on average. The total flight length is 186 km. The raw laser data have been processed into point elevation and 0.5 m grid of the digital elevation model (DEM) by the Jiantong Co., who is a co-investigator of this project. All products are produced according to the criteria of 1:500 scale [18].

The claimed vertical and horizontal uncertainties of the laser point coordinates are less than 0.15 m and 0.25 m, respectively [18]. The quality of the laser point elevation was controlled by using 10 ground control stations that were set up and surveyed before the airplane survey. The inverse distance weighting (IDW) method is applied to interpolate the point elevation into a 0.5 m grid DEM. In July 2018, a set of independent ground elevation data were collected to validate the LiDAR DEM by a precision differential Global Position System (GPS), the Real-Time Kinematic (RTK) instrument (Unistrong G970II). Another suit of levee crown height and cross-section data were surveyed on 26 October 2019 to validate the extracted levee geometric parameters (Figure 1).

All the following analysis is based on the 0.5 m grid data. The first step is to compute the surface slope from the grid elevation data. The slope is defined as an angle made by the horizontal plane and the inclination surface, which is calculated using the neighborhood operation from the 0.5 m grid of LiDAR elevation in a 3 × 3 moving window by Equation (1) in ArcMap 10.4 [19,20].

$$\text{Slope (degree)} = tan^{-1}(((\frac{\partial z}{\partial x})^2 + (\frac{\partial z}{\partial y})^2)^{\frac{1}{2}}) \times \frac{180}{3.14} \quad (1)$$

where ∂_x and ∂_y are the projected runs of the height change along the x and y direction on the x-y plane, and ∂_z is the height change between the central grid versus the 8 neighboring grids. The slope can be expressed as an angle (°) or a ratio of height to run.

2.3. Extraction of Levees Geometric Parameters

Levee is an important water-control engineering measure and plays an important role in flood defense and disaster relief. Figure 2 illustrates the typical shape and elements of a levee section. It has a flat surface (levee crown) and two inclined areas, the waterside and landside. In the coast area, a wave wall is often added on the waterside top of the levee crown to resist wave overtopping. There is a break in the slope between the flat crown/ground surfaces and the inclined areas. This slope break is mainly utilized to extract the levee crown/toe by a slope threshold of 5° according to the levee construction code and the designed criteria [21,22]. A larger range of slopes (0°–8.43°) is also used to classify the levee crown [15]. Obtaining the levee crown/toe is the first critical step to extract the geometric parameters, whose accuracies are determined by the levee crown/toe boundary to some degree. Manual editions are needed to improve the slope-classified levee crown/toe boundary with the help of high resolution images and elevation (Figure 3).

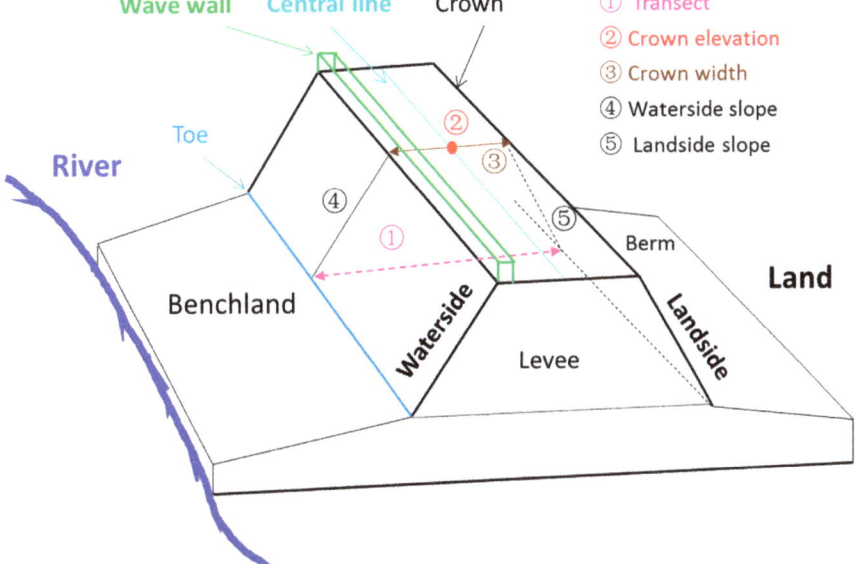

Figure 2. The sketch of the levee elements.

The flowchart and main methods utilized for extracting the levee geometrical parameters are demonstrated in Figure 3. The extracted outlines of the levee crown are used to generate the levee central line, which is further applied to produce a series of transect lines perpendicular to the central line at a 100 m interval using the Thiessen polygons (Figure 4a) [23,24]. Those polygon sides that are perpendicular to the levee central line and constrained by the outlines of levee crown and toes are the levee transect lines (Figure 4b). There are total 317 transect lines in the levee segments from A to D.

Four levee geometric parameters are extracted using those transect lines, including crown crest elevation, crown width and the two slopes of waterside and landside incline faces. The crown elevation is defined as the surface height at the crossing point of the crown central line and the transect line (Figure 2). The length of the transect line constrained by the levee crown outlines is the crown width. The side slope is the mean values of the slope grids over the transect line between the outlines of levee crown and toes each side. The entire processes are automated by Python scripts on the platform of ArcMap 10.4 after the accurate outlines of levee crown and toes are derived.

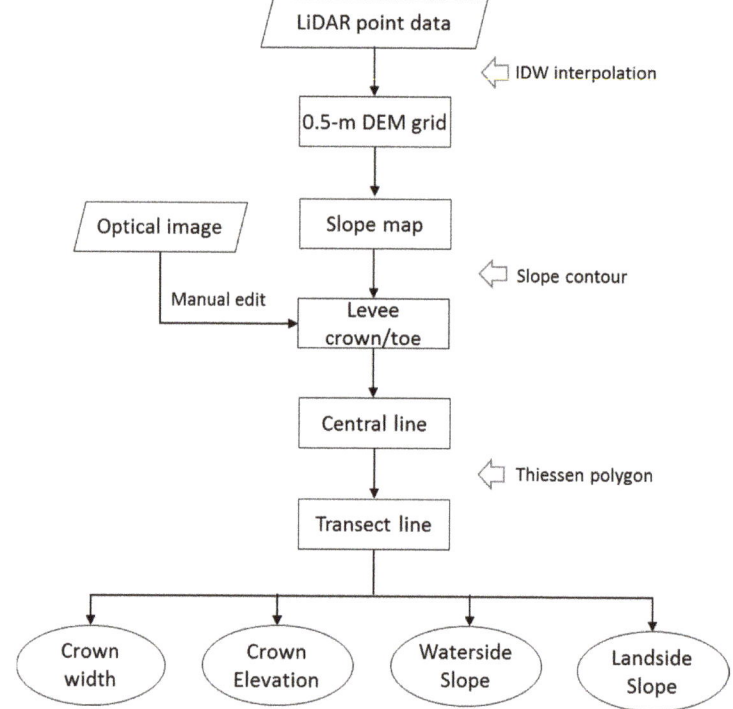

Figure 3. The flowchart and main methods used in extraction of levee geometric parameters.

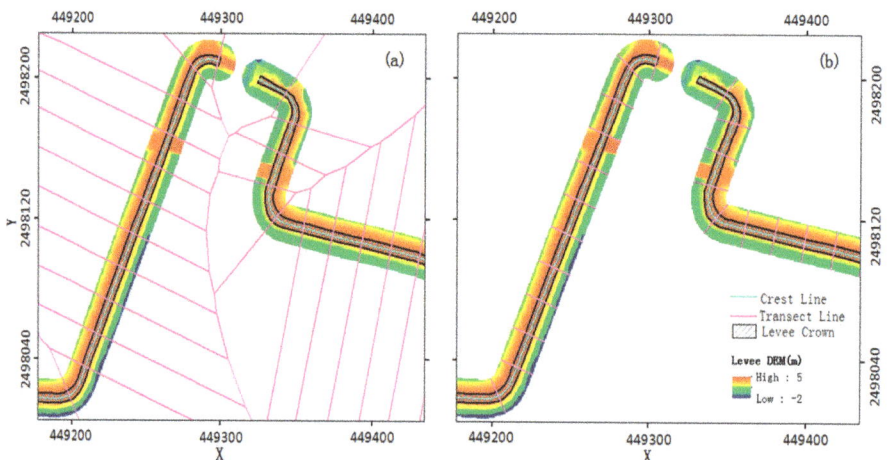

Figure 4. (**a**) Thiessen polygons generated from the levee crown central line at a 100 m interval, and (**b**) the levee transect lines generated by the Thiessen polygons and the outlines of levee crown and toes. The blank area in levee is a sluice for a local channel.

2.4. Elevation Extraction in the Dock Area

The dock in section E (Figure 1) is a transportation hub for passengers onboard and off board, cargo handling, and safe entry and exit of ships. It is also an important protective measure for the safety behind the dock area. There is no standalone levee on the dock area E, behind which a new

Huoju business district has been established during the past two decades. This new business district may face severe flood threat. However, it is difficult to determine the flood-control height since the dock operation area has no regular levee structure. Two surface heights are extracted from the 0.5 m grid elevation data at a 20 m interval. One height is the surface elevation on the dock operation area along the river, and the other is the maximum surface height along a 1-km transect of the dock area. Both surface heights are used to compare with the water levels at different flood frequencies for the assessment of flood defense capacity in the dock area.

3. Methods of Levee Assessment

3.1. Levee Design Criteria

Levees have usually been constructed according to a set of design code of the geometric parameters, such as the crown elevation, crown width, the surface slopes of waterside and landside [7,8,17]. With the age of use, the levee stability and flood-control capability will decline due to sea level rise, flood stage variability, river bed erosion, land subsidence, human disturbance and the functional zone changes behind the levees [16]. It is imperative to frequently assess their functional performance. China issued its latest design code (GB 50286-2013) for the levee construction in 2013 [22]. The design criteria of geometric parameters for the surveyed levees are summarized in Table 1 according to the designed criteria of levee construction/rebuilding and the flood-control level planned in levees A to D. This study uses the airborne LiDAR altimetry-derived geometric parameters to assess whether they can meet the functional requirement according to the latest flood-control planning and the design code (GB 50286-2013). The designed crown elevation (4.5 m) is equivalent to the base height of extreme water level (3.91 m) in a 50-year return period at present and a redundancy height of 0.6 m. There is an additional wave wall on the waterside top of the levee crown for resisting wave overtopping. The height of wave wall varies and is estimated according to the wave conditions. Among the surveyed levees, the wave wall is made of concrete rock/brick and has 20–30 cm width and 30–120 cm height in different segments. There was no wave wall in levee D at present. In addition, the slope criteria are used for earthen dike surface with grass cover and not suitable for concrete inclined surface with berm. The design criteria are relatively low compared to those adopted in the International Levee Handbook [25].

Table 1. The design criteria of geometric parameters for soil levees A to D.

Levee Section	Engineering Grade	Crown Elevation (m)	Crown Width (m)	Waterside Slope (°)	Landside Slope (°)
A	III	≥4.5	≥6	≤1:2	≤1:2 (26.6°)
B	II	≥4.5	≥6	≤1:2	≤1:2
C	III	≥4.5	≥4	≤1:2	≤1:2
D	III	≥4.5	≥4	≤1:2	≤1:2

3.2. Calculation of Extreme Water Levels

The hydrodynamic environment has undergone fast changes in the Hengmen waterway during the past decades. The frequency of the tidal levels or water levels has been changing as well [26,27]. In order to assess the actual capacity of levees flood defense, the extreme water levels are calculated by the Gumbel method using all the latest water level records. The probability density and cumulative distribution function are expressed by Equations (2) and (3). Equation (4) is used to calculate the extreme water level at a return period (T year) [28].

$$f(x) = \frac{1}{\alpha} \exp\{-[(x-\beta)/\alpha]\}^{-\exp\{-[(x-\beta)/\alpha]\}} \qquad (2)$$

$$F(x) = e^{-e^{-[\frac{x-\beta}{\alpha}]}} \qquad (3)$$

$$X_T = \beta - \alpha \ln\{-\ln[1 - (1/T)]\} \quad (4)$$

where x is the calculated variable (water level), α and β are the estimated parameters, X_T is the water level at a return period (T year).

According to the annual maximum water levels recorded from 1958 to 2018 (missing data at some years) at the Hengmen tidal station (Figure 1), the computed water levels for the four return periods (T) of 20, 50, 100, 200 years are 3.54, 3.91, 4.19 and 4.47 m (1985 Yellow Sea Geodatum), respectively.

The China Sea Level Bulletin in 2017 reported that the mean sea level rise rate was 3.3 mm/year from 1980 to 2017 along the coast of China [29]. It is expected a range of 65–170 mm sea level rise in the coming 30 year in the coast of Guangdong Province, where the Pearl River estuary of this study is expected to have the largest rate. Therefore, an upper-end sea level rise scenario (0.5 m) is considered in assessing levee's flood defense capacity at the end of this century [30,31].

3.3. Assessment of Levee Stability and Overtopping Threats

One design standard is usually applied to assess levee stability and overtopping risk in previous studies [7,9,16], which did not provide the specific information for those below the design standard, especially for the crown height. This information is important for flood risk assessment and the levee improvement planning. This study proposes a scoring approach to assess levee stability and overtopping threats based on the criteria of levee geometric parameters and the extreme water levels at four flood stages or frequencies of 1:20, 1:50, 1:100, and 1:200 years. The full score is 10 points with maximum 4 points for the crown height and 2 points for crown width, waterside slope, and landside slope, respectively (Table 2). The scores of the crown height are assigned by comparing the crown height to the extreme water levels at four flood frequencies. For instance, if the crown height is less than 3.5 m corresponding to a return period of 20 year, its score is zero, and those higher than the designed elevation of 4.5 m (T = 200 years) get 4 points. The criteria for the crown width are 6 m for levees A and B and 4 m for others. If the crown width is larger than or equal to 6 m for levees A and B(4 m for C and D), it gets the full 2 points, 1 point for 3–6 m (3–4 m for C and D), and zero for those less than 3 m. The maximum standard for the slope are 26.6° for all levees A, B, C, and D. If the slope is smaller than or equal to 26.6°, it gets the full 2 points, 1 point for 26.6°–33.7°, and zero for larger than 33.7°. The assessment scores are normalized into percentage in order to compare with the common qualified rates, which compute the percentage of parameters equal to or above the designed standards [7]. In addition, on the waterside, the levee toe is paved with rock revetment to withstand wave erosion. A concrete berm is often built above the levee toe. The waterside concrete surface is quite steep, and its side slope is not assessed since the slope criteria are used for earth dike and not suitable for concrete inclined surface with berm on the waterside.

Table 2. The scoring sheet of levee stability and overtopping risk. The crown height intervals are set according to the flood stages (water levels) at four return periods.

Assessment Score	Crown Height (m)	Return Period (Year)	Water Level (m)	Crown Width (m)		Landside Slope (°) A,B,C,D
				A,B	C,D	
0	<3.5	20	3.54	<3	<3	>33.7 (1:1.5)
1	3.5–3.9	50	3.91	3–6	3–4	26.6–33.7
2	3.9–4.2	100	4.19	≥6	≥4	≤26.6 (1:2.0)
3	4.2–4.5	200	4.47			
4	≥4.5					

4. Results

4.1. Validation of Geometric Parameters

Figure 5 illustrates the original LiDAR points cloud elevation, high resolution optical image, the produced 3D real scene model, and the interpolated 0.5 m DEM. Besides the trees and vegetation, the detailed 3D variation of levee can be revealed by the high precision LiDAR points compared to the optical image, for instance, the bulges of grass on the landside and wave wall on the waterside (Figure 5a,b). Combination of both LiDAR point elevation and optical image reproduces the levee real situations by the levee 3D real scene model, which is useful in levee management and instability examination.

The accuracies of LiDAR elevation and the extracted levee geometric parameters are validated using two sets of ground survey data using the RTK instrument (Figures 6 and 7). The locations of the surveyed sites are illustrated in Figure 1. Both the LiDAR elevation and RTK measured ground elevation have a good linear relationship ($R^2 = 0.97$) and a Root Mean Squared Difference (RMSD) of 0.10 m (Figure 6a), which falls within the vertical uncertainty range (<0.15 m) of the laser system [18]. In contrast, the extracted levee crest heights along the levee crown central line even have a smaller discrepancy (RMSD = 0.05 m) with the RTK measured elevation (Figure 6b).

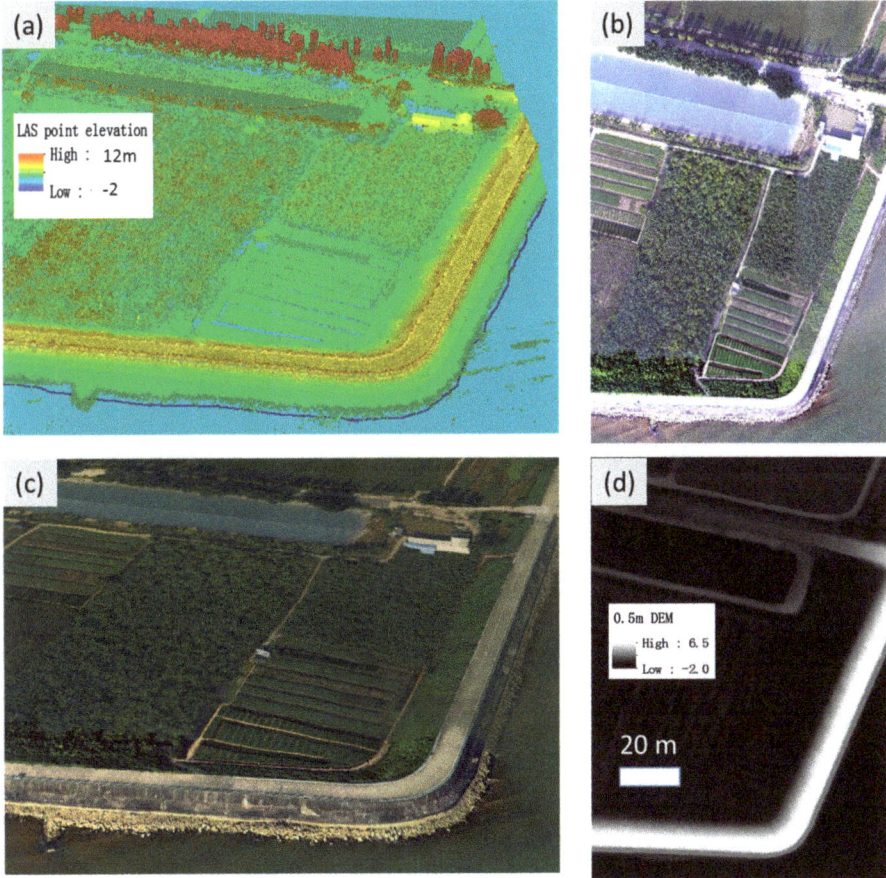

Figure 5. Demonstrations of (**a**) the original LiDAR point cloud elevation (16 points/m^2), (**b**) 0.05 m optical image, (**c**) levee 3D real scene model, and (**d**) 0.5 m DEM.

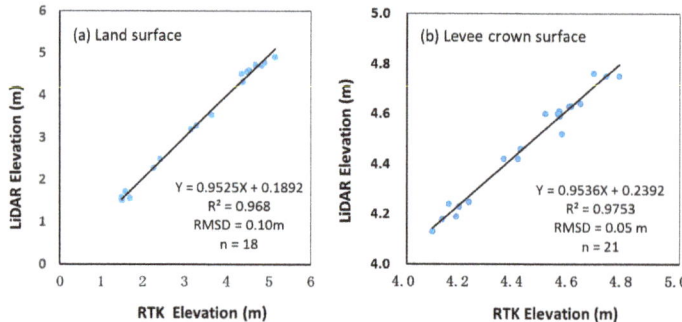

Figure 6. Comparison of ground measured elevation using GNSS RTK and the airborne LiDAR for (**a**) the original 0.5 m grid DEM on land surface and (**b**) the levee crown height extracted from the 0.5 m DEM by the script.

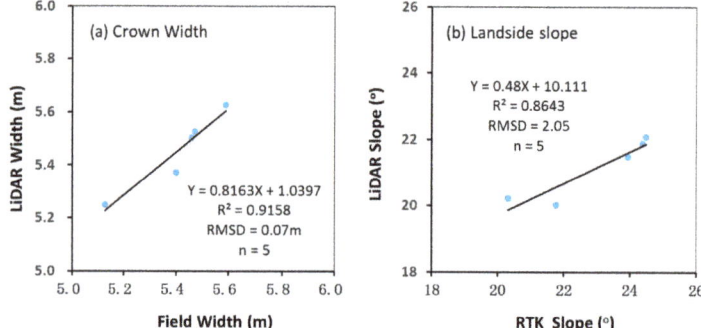

Figure 7. Scatter plots of the script-extracted levee crown width and landside slope against the field measured values using meter cord and RTK in levee A.

The extracted crown widths also have a good agreement (RMSD = 0.07 m) with the field measurements (Figure 7a). The mean slope extracted from the LiDAR data is about 2° smaller than the field measurements (Figure 7b). The slope discrepancy is mainly attributed to the dense vegetation (grass) on the landside inclined face (Figure 5), where the LiDAR only can measure the surface elevation, but the RTK surveys the land surface under the grass. Their mean elevation difference is −0.28 m (LiDAR-RTK) on the five levee cross-sections of the inclined landside faces. The mean slope of levee waterside is hard to validate and not included in the analysis since it contains a berm and rock-paved levee toe (Figure 5c).

4.2. Levee Stability Assessment

Levee morphometric modifications are often caused by crest subsidence, crown narrowing and slope steepen, and affect the levee integrity and working performance [7]. Table 3 summarizes the mean, standard deviation, qualified rates and the assessment scores of the levee geometric parameters. The assessment scores are higher than the qualified rates since those lower than their standards still win some credits (Table 3). Overall, the crown elevation has the lowest assessment scores especially for levee C and E. The landside slopes have the highest assessment scores and near all are qualified. The waterside slopes are not assessed since they are concrete surface with berms and the levee toe is paved with rock revetment for resisting wave erosion. There was no suitable national design standard for them. Levees A and B have higher assessment scores than levees C, D and E.

Levee A is the southern part of the Minsan joint levee, which protects a population of over 250, 000 and an area near 190 km². The levee geometric parameters should meet the designed standards (4.50 m) for a flood level (3.91 m) at the frequency of 1:50 year and a 0.6 m redundancy height (Table 1). The total assessment score is 85% for the three geometric parameters among the 125 transect lines (Table 3). The mean crown crest height is 4.70 ± 0.46 m (standard deviation), and 90% (qualified rate) of them are equal to or higher than the 4.5m standard. The assessment score declines from 96% to 65% when a 0.5 m sea level rise is considered. The qualified rate is only 7%, and the assessment score is 53% for the crown width. The qualified rate and assessment scores of landside slope are 90% and 91%, respectively. As shown in Figure 8, the low-score section is mainly contributed by low crown elevation in the west of levee A and marked as ①, where all crown heights are lower than 4.5 m within the 1.7-km levee (17 transects).

Table 3. Statistic of levee geometric parameters and assessment results. The qualified rate is defined as the count percentage that each parameter meets the designed standards. The assessment score (AS) is normalized into percentage by comparing the actual scores to the full scores for each and all four parameters within each levee segment. A 0.5 m Sea Level Rise (SLR) is also considered.

Levee Sections		A	B	C	D	E
Count	levee transect lines	125	35	47	110	266
Crown elevation	Mean (m)	4.70	4.75	3.98	4.34	4.00
	Standard deviation	0.46	0.81	0.71	0.51	0.55
	Qualified rate	90%	89%	9%	27%	26%
	Assessment score	96%	97%	48%	72%	48%
	AS after 0.5m SLR	65%	70%	18%	35%	24%
Crown width	Mean (m)	5.38	6.90	4.52	4.49	-
	Standard deviation	0.78	1.19	0.98	0.65	-
	Qualified rate	7%	100%	77%	89%	-
	Assessment score	53%	100%	88%	94%	-
Slope landside	Mean (°)	18.9	15.2	7.8	7.1	-
	Standard deviation	7.4	6	6.7	7	-
	Qualified rate	90%	100%	98%	97%	-
	Assessment score	94%	100%	98%	98%	-
All	Assessment score	85%	99%	71%	84%	48%

Levee B is a small eastern section (~3.5 km) of the Zhongshun joint levee, which protects over 640,000 people and areas of 700 km². It is classified as national grade II levee and primarily escorts the flood water of Xijiang to the bay of Lingdingyang through the Xiaolan-Hengmen waterway. It has the highest assessment score (99%) among the four segments from A to D (Tables 1–3, Figure 8). Both the crown width and landside slope have the 100% qualified rates and assessment scores. 89% of the crown heights are higher than 4.5 m, and only 4 heights are slightly less than 4.5 m but higher than 3.91 m (minimum height of 4.34 m). In addition, there is a concrete wave wall of about 1 m high and 0.2 m wide on the levee crown. It is not accounted in the levee crest height. The assessment score of the crown height declines from 97% to 70% when the sea level rises by 0.5 m, while it rises to 100% when the wave wall height is added to the levee crown height.

Levee C is located in the southern bank of the Hengmen waterway and protects the eastern part of the new Huoju business district. Its design standard recently has been modified to the same criteria as levee A except for the crown width. Both the crown width and slopes have high assessment scores. The crown heights have an assessment score of 48%, and only 4 heights (9%) are higher than 4.5 m among the total 47 transects (Table 3). The eastern section of levee C has been improved to meet the design standard, but the most western part failed to meet the standard (Figure 8).

Levee D is the northern part of the lately planned Cuiheng high-tech district, which was reclaimed from the intertidal beach and is being improved to resist a 1:50-year storm surge and tidal level.

At present, the mean crown height is 4.34 ± 0.51 m, and only 27% of them could meet or higher than the 4.5 m standard, plus 39% within 4.2–4.5 m and 27% within 3.9–4.2 m (Table 3, Figure 8). The eastern part that faces the Lingdingyang has higher elevation up to 5.35 m to resist the storm surges. Some crown surfaces were planted with bananas. Ground surveys also identified levee washing out and several collapses. Now the levees are constructing with updated design criteria, and the old levee that was surveyed in December 2016 had been partially reinforced in 2019.

Figure 8. Spatial distribution of assessment results. The normalized assessment score is used for levees A, B, C and D. The water levels at the four return periods are illustrated to compare with the dock front elevation along the river of levee E. The number of ① to ④ represents the locations of levee crown modifications and notches.

4.3. Levee Overtopping Assessment

The above stability assessment and the assessment scores can reveal whether the levee meets the corresponding designed standards and the overall stability or working conditions. However, these parameters are extracted by transect lines at an interval of 100 m, and the discrete crown heights often miss some overtopping threat from local disturbances [7]. Therefore, the continuous crown elevations are extracted to assess the levee overtopping threat from the 0.5 m grid LiDAR DEM by comparing to the designed elevation and the extreme water levels in four return periods of 20, 50, 100 and 200 years (Figure 9).

The designed elevation coincidentally matches the extreme water level in a 200-year return period for the surveyed levees (Figure 9). As demonstrated by the assessment scores in Table 3, levees A and B have much higher crown elevation than levees C and D. Levee A has some overtopping threats due to levee crown modifications although its main crown elevation is above the extreme water level in the 200-year return period (Figure 9a). There is about 300 m levee with abnormally lower crown elevation marked as ① in the west of levee A (Figure 9a). It is up to 30 cm lower than the neighboring levee crown height (Figure 10a). Ground surveys identify that the levee top (crown height + wave wall) has relatively consistent heights of ~4.8 m along this section in spite of its lower crown surface. In other words, the lower crown surface is compensated by adding the wave wall to the levee crown for flood defense, and was intentionally constructed to reduce the levee weight over the soft sediment layer beneath the levee.

The levee A has several other notches, which bring additional overtopping threats (Figure 9a). Ground surveys on 12 July 2018 and 26 October 2019 identified that those notches were mainly caused by local human activities. For instance, the notches ② and ③ were up to 1.4 m and 1.9 m lower than the levee top (levee crown elevation plus 30-cm wave wall height), respectively (Figure 10b,c). The notch ② actually has two notches, which were modified for traffic crossover for a local steel pipe factory dock and a ferry crossing the Hengmen waterway between the town of Minzhong and the city of Zhongshan (Figure 10b). Ground survey also found that the water level rose to 3.5 m nearby the ferry by wave setup on 22 August 2017 after the 1713 typhoon Hato made landfall about 60 km away in the southwest estuary. It would overtop the western notch ② with a surface height of only 3.40 m. Behind the levee, the street elevation of the village is just 1.6 m and would be inundated for near 2 m deep if the notch had not been blocked by local villagers with sand bags. The notch ③ was modified for a shipyard built within the levee benchland (Figure 10c). In front of the shipyard, the surface elevation of the levee notch is only 3.26 m, which is even less than the water level (3.54 m) in a 20-year return period. There are 24-hour gate guards in the shipyard, and the notch can be quickly filled by the woodblock and sand bags ahead of flood (Figure 10c). However, other notches like ② without human guard should be paid more attention. Ground survey also found many other small notches on the levee wave wall for the access convenience of local activities. Therefore, it is imperative to timely assess the levee and repair those cutoffs to reduce the flood overtopping threats.

The crown elevation of levee B is in the best condition among the surveyed levees, plus a 1m height wave wall on the waterside levee crown (Figures 8 and 9b). The overtopping threat of levee B is relatively low. Even after the sea level rises by 0.5 m, the crown elevations of levees A and B are still higher than the extreme water level (4.41 m) in a 50-year return period (Figure 11a,b). When the wave wall height is added to the levee crown height, the levee top height is still higher than the extreme water level (5.0 m) even if the sea level rises by 0.5 m (Figure 11). However, there is a 300 m section from 2900th to 3200th m where the crown height is lower than the designed elevation of 4.5 m and has a minimum height of 4.34 m. This lower section is likely related to levee subsidence that is up to 0.2 m compared to the neighboring levee crown elevation (Figure 9b, inset plot). This subsidence occurs in the western Hengmen waterway where the Jiya and Xiaolan waterways are confluent (just in the west of the bridge) (Figure 8).

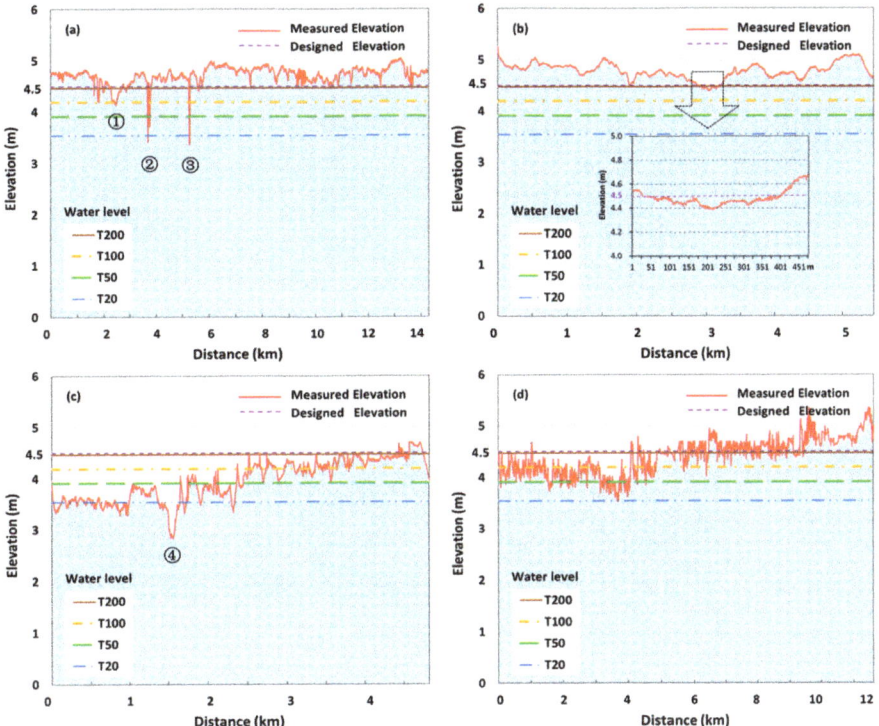

Figure 9. Comparison of the continuous crown elevation from the 0.5 m grid DEM against the designed elevation and the extreme water levels in the four return periods of 20, 50, 100 and 200 years for levees A (**a**), B (**b**), C (**c**) and D (**d**). The distance starts from the west within each levee.

The 5-km levee C is divided into five segments by four sluice gates and local short open channels. Each block separated by the channels represents a typical functional zone, such as waste water treatment plant, sand dock, fishing ponds (filled up), gas dock and village from west to east in sequence (Figure 8). The crown elevation of the eastern half levee is above the extreme water levels (3.91 or 4.19 m) in a 50 or 100-year return period and is much higher than the western part, whose elevation is just around the extreme water level (3.54 m) in a 20-year return period except for the section ④ (Figure 9c). Like the section ① in levee A, this section ④ locates in the sand dock area, where levee crown surface elevation was up to 0.6 m lower than its nearby levee surface elevation (Figure 10d). In spite of the lower levee surface elevation, there is a 1.2 m height wave wall, which increases the levee top to 4.0 m and makes it higher than the extreme water level (3.91 m) in a 50-year return period. Meanwhile, the levee C has been planned to meet the same design standards of levee A to protect the new Huoju business district. Particularly, the waste water treatment plant in the west needs higher levee to protect from the flood water although it is separated from other sections. Once it is flooded, severe environmental pollution may occur besides damages to the plant.

The crown surface was quite rough in levee D in the new Cuiheng high-tech district since most of them had not been concreted (Figures 8 and 9d). The crown heights in the eastern outer part facing the bay were higher than the 4.5 m standards, while most crown heights in the west inner part were in a range of 4.0–4.5 m and a small part was lower than 4.0 m. Most part of levee D was irregular and even weak, and part of the eastern crown surface were planted with bananas. New constructions are carrying out to improve the levee to protect the perspective dense population and high-value infrastructures in this high-tech district.

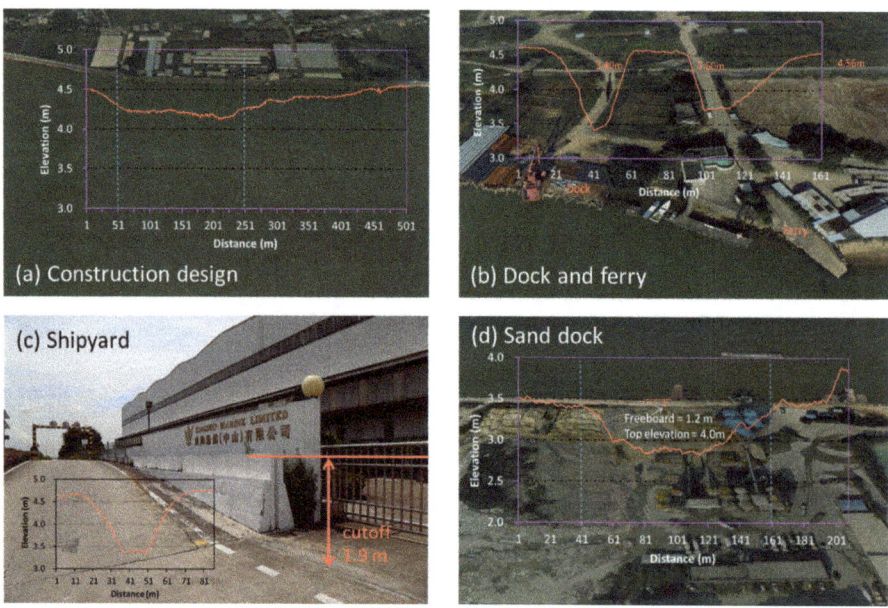

Figure 10. Examples of levee modifications and traffic notches of (**a**) construction design, (**b**) dock and ferry, (**c**) shipyard in levee A, and (**d**) sand dock in levee C. The inset plots are the surface elevation over the levee modification marked as ① to ④ in Figures 8 and 9. The plot range is adjusted to match the horizontal size of levee modification in plots (a), (b) and (d).

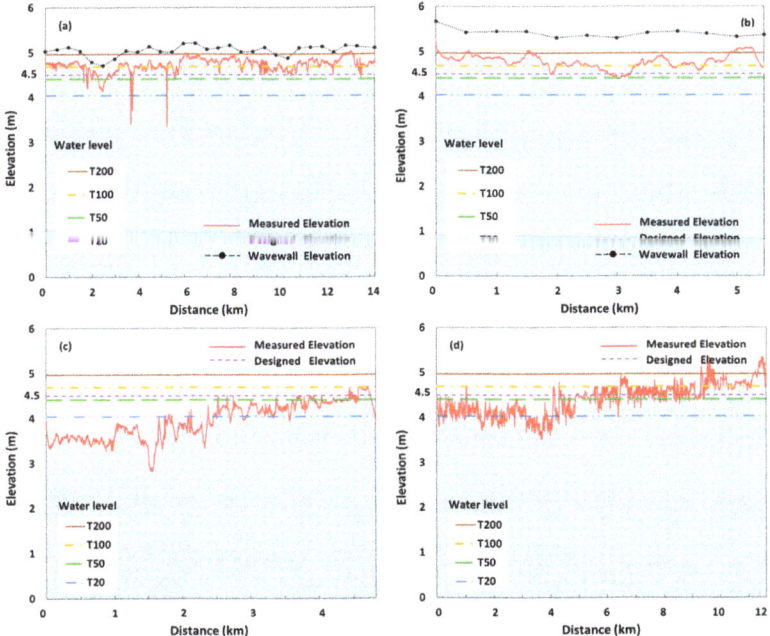

Figure 11. Comparison of the continuous crown elevation from the 0.5 m grid DEM against the designed elevation and the extreme water levels in the four return periods of 20, 50, 100, and 200 years when the sea level rises by 0.5 m for levees A (**a**), B (**b**), C (**c**) and D (**d**). The wave wall elevation (wave wall height plus levee crown elevation) is illustrated as the levee top height in levee A and B.

Levees E and C together protect the new Huoju business district (~20 km²) of Zhongshan. The levee E starts from the Donghe sluice gate in the west and connects levee C at the Xiaoyin sluice gate in the east. It protects an area about 12 km². The river front area has been commercialized for sand dock, cruises harbor, container dock, petrochemical plant, pharmaceutical and condiment factories from the west to east in sequence (Figure 8). There is no regular levee in this area, such as the levee crown or levee width. The levee E is like a sea wall with near-vertical and concrete waterside surface. The river front elevations are primarily used to compare with the extreme water levels in the four return periods. The front elevation can be divided into three sections (Figure 12a). The western 2-km section is the sand dock with most front elevation around the extreme water level (4.19 m) in a 100-year return period. The eastern 1.6-km section is seated with several large companies and its front elevation is above the designed elevation of 4.5 m. Even after the sea level rises by 0.5 m, most front elevations are still around the extreme water level (4.69) in a 100-year return period (Figure 12b). The middle 3-km section is the cruises harbor and container dock with front elevation slightly below the extreme water level (3.54 m) in a 20-year return period. The primary purpose of the low elevation is designed for dock operations, instead of flood defense. An alternative is to search the maximum elevation along a 1-km transect line perpendicular to the river front as a backup for flood defense. The max elevation is generally above the extreme water level (3.54 m) in a 20-year return period, and part of them is above the level in a 50-year return period. Part of the transect lines are just 150 m wide and do not cover the max elevation because of the narrow LiDAR coverage in this area. Anyway, it is urgent to add a backup levee behind the dock operation area to protect the fast-developing Huoju business district.

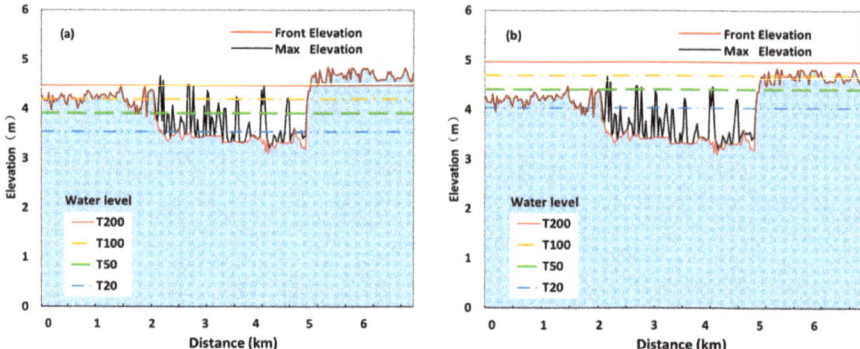

Figure 12. Comparison of the crown elevation of levee E (dock area) from the 0.5 m grid DEM against the extreme water levels in the four return periods of 20, 50, 100, and 200 years (**a**) at current sea level and (**b**) after a sea level rise of 0.5 m. The distance starts from the west.

5. Discussions

It is not a trivial task to accurately extract levee geometric parameter from the LiDAR elevation data although it is just a two-hour flight data. One of the critical steps is to derive the crown outline, which is the most crucial geometric parameter for automatically extracting other geometric parameters. For the standard levees such as those in levees A and B, where the crown surface is concrete and has little trees or vegetation, both the slope contour or image classification methods work well in extracting the crown outlines. However, the field conditions of levees are complex, such as in the commercialized river bank in levee E and the irregular crown surface partially planted with bananas in levee D. No single method can deal with all conditions. All DEM, slope and high resolution images (0.05 m) are used together to extract the crown outlines, and much manual edition are still needed [15,16]. Subsequently, other geometric parameters can be automatically extracted using the procedure developed in this study, which greatly improves the speed of data processing. Meanwhile, manual check is also needed for those abnormal low or high values with the help of the original LiDAR points cloud elevation and the levee 3D real scene model built in this project (Figure 5). A good case in point is the levee crest heights that are extracted over the central line. Some sections of the central line are not on the crown crest since local traffic connection often cuts part of the crown surface, leading to abnormal lower elevation (Figures 9 and 10).

The geometric parameters extracted by the automated script have good agreement with ground measurement, especially on the levee crown surface (Figures 6 and 7). However, the number of in-situ slope profiles and crown width is limited, and more ground surveys are needed to confirm the reliability of the script in future application. Moreover, since there is some time lag between LiDAR surveying time (December 2016) and GPS-RTK in situ measurements (July 2018 and October 2019), the RTK survey sites were selected over relative stable levee section, thus mitigating the effect of levee modification with time. The results also shown that the levee crown surface heights surveyed with RTK in October 2019 have good agreement with the LiDAR ones surveyed in December 2016 (Figure 6).

The normalized assessment scores for levee stability designed in this study are higher than the qualified rates used in previous studies [7,9,16]. The primary difference is attributed to the multiple credits for those that do not meet the standard, which were given a zero credit in previous studies (Tables 1–3). Thus the assessment scores offer more information for levee actual conditions and are superior to the simple qualified rates, which only consider the credit of parameters meeting the designed standards and ignore those even they are just slightly lower than the designed standards.

The assessment score provides an overall view of levee stability, i.e., the morphological modifications (Table 3, Figure 8), which mainly embody in slopes and crown elevation. The continuous parameters such as the crown elevation give more information than those extracted over the transect line at a certain interval. For instance, levee crown modifications are observed in all levees by checking the continuous levee crest height (Figure 9). Field survey, high resolution images and levees 3D real scene model are also needed together to identify what causes and how large are the levee morphological changes, such as levee surface subsidence due to natural subsidence in levee B, the combined effect of natural sink and industry operation in levee A and C, artificial cutoffs for traffic crossover in levee A (Figures 9 and 10). The information is crucial to offer suitable remediation options, especially in the flood season and for emergency response. However, this study is a generic assessment of levee stability and overtopping threats by only using the levee geometric parameters. It examines whether the levee has any distortion, breach or slides occurred, rather than to investigate why and what cause the instability. More investigations are needed to reveal the mechanisms of levee instability caused by different factors, such as the natural subsidence in levee B due to compaction, river scoring and groundwater seepage etc., and those impacted by human activities in other levees [25].

For the overtopping threats, we compare the levee crown heights with the extreme water levels estimated from the water level records at one tidal station (Figure 1). This is a gross assessment because of the changing hydrodynamic environment and the water level gradient due to coastal storm surge. The time series of the water level data may be non-stationary due to the changing hydrodynamic environment. To reduce this type of uncertainty, we do not use the old design water levels applied in levee construction, but using those estimated from all data including the latest water level records in spite of its non-stationarity. It is common and the only choice at most conditions to use historic data to predict current and future flood peak magnitudes assumes that the historic data is representative of the present and future conditions at the project site [25].

On the other hand, representation of extreme water levels using one station at a larger area brings another type of uncertainty in this study. The surveyed levees span near 20 km from west to east, and the tidal station is located in the middle section (Figure 1). However, there is only one tidal station that has long-term (>20 years) water level records suitable for extreme water level analysis in different frequency/return periods. Our investigation from recent in situ observation and hydraulic modeling find that the gradient of peak water levels is 0.5 m per 10 km in the surveyed levees when a 1:50-year flood discharge from the upstream watershed encounters with a 1:50-year coastal storm surge. The gradient is mostly affected by the scale of storm surges. The coastal sections face higher water levels due to storm surge and wave setup. This indicates that levees in the coast part would face higher overtopping threats and should have higher design standards than the west. Our results show that the ocean side of levee D does have higher crown heights than the inner river side, but levee A has a near-unanimous crown top height along the 12.5-km length from east to west. A higher levee crown height and wave wall should be considered in the future coastal levee reinforcement.

6. Summary

The coastal cities in the low-lying Pearl River Delta face severe flood threat although they are under the protection of a huge levee system. Once a flood event happens, it often causes catastrophic impacts on property and life loss behind the areas protected by levee as more and more population and properties move in. Routine and fast assessment of levee is critical to guarantee the community safety surrounded by levees. This study designs a scoring approach to quickly assess levee stability and overtopping threats with geometric parameters derived from the high-precision airborne LiDAR data. The procedures and assessing approach developed in this study can be easily applied for the levees assessment in the entire Pearl River Delta and somewhere else in the world.

The airborne LiDAR and high resolution images enable fast and large scale examination of the levee physical conditions. However, it is not a trivial task to accurately extract levee geometric parameters from the LiDAR elevation data. This study developed a procedure to automatically extract the levee geometric parameters for levee assessment, such as the crown elevation, crown width, waterside slope and landside slope. Meanwhile, manual edition and quality check are still needed especially for the most critical crown outlines and the crown crest heights.

This study designs a scoring approach to assess the levee stability and overtopping threats with levee geometric parameters. The normalized assessment scores are higher than and superior to the qualified rates used in previous studies. This is because the scoring approach compares the geometric parameters to several standards and assigns multiple credits for those that do not meet the designed standards. In contrast, they are given a zero credit in previous studies. Levee A and B have much higher crown elevation and assessment scores than others. Their crown crest heights are above the flood level (4.5 m) in a 200-year return period and can still be above the flood level in a 50-year return period even if the sea level rises by 0.5 m. However, the continuous crown heights reveal several levee morphological modifications in all levees surveyed, including the best levees A and B. Those modifications are primarily caused by natural subsidence in levee B, the combined effect of natural sink, industry operation and even special construction design in levee A and C, and artificial cutoff for traffic crossover in levee A.

The geometry of levee D is not as regular as levee A or B. Their standards had been planned to the same as levee A to protect the new Cuiheng high-tech district. The crown heights in the eastern outer part were higher than the 4.5 m (designed standards), while most crown heights in the west inner part were in a range of 4.0–4.5 m and a small part was lower than 4.0.

Levees E and C together protect the new Huoju business district (~20 km^2). Their standards had also been planned to the same as levee A. The crown elevation of the eastern half levee C is above the extreme water levels (3.91 or 4.19 m) in a 50 or 100-year return period, while it is just around the flood level (3.54 m) in a 20-year return period in the western part. The river front area in levee E has been commercialized and there is no regular shape. The river front elevations in the western 2-km section are around the flood level (4.19 m) in a 100-year return period, and they are above the designed elevation of 4.5 m in the eastern 1.6-km section. The middle 3-km section is the cruises harbor and container dock with front elevation slightly below the extreme water level (3.54 m) in a 20-year return period. The maximum elevation along a 1-km transect line is generally above the flood level (3.54 m) in a 20-year return period. It is urgent to add a backup levee behind the dock operation area to protect the fast-developing Huoju business district.

Author Contributions: Conceptualization, X.W. and T.Z.; methodology, X.W. and L.W.; software, X.W. and T.Z.; validation, X.W., L.W. and T.Z.; formal analysis, X.W. and L.W.; investigation, X.W. and L.W.; resources, X.W. and T.Z.; data curation, X.W., L.W. and T.Z.; writing—original draft preparation, X.W.; writing—review and editing, X.W.; visualization, L.W.; supervision, X.W. and T.Z.; project administration, X.W. and T.Z.; funding acquisition, X.W. and T.Z. All authors have read and agreed to the published version of the manuscript.

Funding: This study is funded by the National Natural Science Foundation of China (#41871085) and the Water Resource Science and Technology Innovation Program of Guangdong Province (#2016-19).

Acknowledgments: The comments and suggestions from anonymous reviewers greatly improve this manuscript and are highly appreciated.

Conflicts of Interest: The authors declare no conflict of interest.

References

1. Wang, X.N.; Wang, X.; Zhai, J.; Li, X.; Huang, H.; Li, C.; Zheng, J.; Sun, H. Improvement to flooding risk assessment of storm surges by residual interpolation in the coastal areas of Guangdong Province, China. *Quat. Int.* **2017**, *453*, 1–14. [CrossRef]
2. Florsheim, J.L.; Dettinger, M.D. Climate and floods still govern California levee breaks. *Geophys. Res. Lett.* **2007**, *34*, 22403. [CrossRef]
3. Gallegos, H.A.; Schubert, J.E.; Sanders, B.F. Two-dimensional, high-resolution modeling of urban dam-break flooding: A case study of Baldwin Hills, California. *Adv. Water Resour.* **2009**, *32*, 1323–1335. [CrossRef]
4. Hanson, S.; Nicholls, R.; Ranger, N.; Hallegatte, S.; Corfee-Morlot, J.; Herweijer, C.; Chateau, J. A global ranking of port cities with high exposure to climate extremes. *Climatic Change* **2011**, *104*, 89–111. [CrossRef]
5. Burton, C.; Cutter, S.L. Levee Failures and Social Vulnerability in the Sacramento-San Joaquin Delta Area, California. *Nat. Hazards Rev.* **2008**, *9*, 136–149. [CrossRef]
6. Tobin, G.A. The levee love affair: A stormy relationship? *Water Resour. Bull.* **1995**, *31*, 359–367. [CrossRef]

7. Casas, A.; Riano, D.; Greenberg, J.; Ustin, S. Assessing levee stability with geometric parameters derived from airborne LiDAR. *Remote. Sens. Environ.* **2012**, *117*, 281–288. [CrossRef]
8. National Research Council (NRC). *Levee policy for the National Flood Insurance Program*; National Academy Press: Washington, DC, USA, 1982.
9. Choung, Y. Mapping risk of levee overtopping using LiDAR data: A case study in Nakdong River Basins, South Korea. *KSCE J. Civ. Eng.* **2015**, *19*, 385–391. [CrossRef]
10. Hossain, A.K.M.A.; Easson, G.; Hasan, K. Detection of Levee Slides Using Commercially Available Remotely Sensed Data. *Environ. Eng. Geosci.* **2006**, *12*, 235–246. [CrossRef]
11. Neuner, J.A. Detection of surficial failures in high plasticity, compacted clay slopes using remote sensing along the Mississippi River Levee. Master's Thesis, University of Mississippi, Oxford, MS, USA, 2002; 131p.
12. Kuszmaul, J.; Neuner, J.; Hossain, A.; Easson, G. The use of multispectral imagery to detect variations in soil moisture associated shallow soil slumps. *EOS Trans. AGU* **2004**, *85*(17).
13. Hossain, A.K.A.; Easson, G. Predicting shallow surficial failures in the Mississippi River levee system using airborne hyperspectral imagery. *Geomatics, Nat. Hazards Risk* **2012**, *3*, 55–78. [CrossRef]
14. Bishop, M.J.; McGill, T.E.; Taylor, S.R. Processing of laser radar data for the extraction of an along-the-levee-crown elevation profile for levee remediation studies. *Defense and Security* **2004**, *5412*, 354–359.
15. Choung, Y. Mapping Levees Using LiDAR Data and Multispectral Orthoimages in the Nakdong River Basins, South Korea. *Remote. Sens.* **2014**, *6*, 8696–8717. [CrossRef]
16. Palaseanu-Lovejoy, M.; Thatcher, C.A.; Barras, J.A. Levee crest elevation profiles derived from airborne lidar-based high resolution digital elevation models in south Louisiana. *ISPRS J. Photogramm. Remote. Sens.* **2014**, *91*, 114–126. [CrossRef]
17. United States Army Corps of Engineers (USACE). *Geotechnical levee practice standards operating procedures (SOP) Sacramento District*; United States Army Corps of Engineers (USACE): Folsom, CA, USA, 2008.
18. Jiantong Co. *Technical Report on Airborne Light Detection And Ranging (LiDAR) project of Zhongshan Hengmen Waterway*; Internal Document: Guangzhou, Guangdong, China, 2016.
19. Srinivasan, R.; Engel, B.A. Effect of Slope Prediction Methods on Slope and Erosion Estimates. *Appl. Eng. Agric.* **1991**, *7*, 779–783. [CrossRef]
20. Burrough, P.A.; McDonell, R.A. *Principles of Geographical Information Systems*; Oxford University Press: New York, NY, USA, 1998; 190p.
21. Choung, Y. Accuracy assessment of the levee lines generated using lidar data acquired in the Nakdong River basins, South Korea. *Remote. Sens. Lett.* **2014**, *5*, 853–861. [CrossRef]
22. Ministry of Water Resources of the People's Republic of China, *Code for design of levee (GB 50286-2013)*; China Planning Press: Beijing, China, 2013.
23. Skare, Ø.; Møller, J.; Jensen, E.B.V. Bayesian analysis of spatial point processes in the neighbourhood of Voronoi networks. *Stat. Comput.* **2007**, *17*, 369–379. [CrossRef]
24. Shen, D. Study on flood risk assessment technology and method of Dongting Lake flood storage and detention basin based on airborne LiDAR data. Doctoral Dissertation, NanJing University, Nanjing, China, 2017.
25. Hemert, H.V.; Igigabel, M.; Pohl, R.; Sharp, M.; Simm, J.; Tourment, R.; Wallis, M. The International Levee Handbook. 2013, published by CIRIA, Griffin Court, 15 Long Lane, London, EC1A 9PN, UK. Available online: https://www.ciria.org/ciria/Resources/Free_publications/I_L_H/ILH_resources.aspx (accessed on 8 January 2020).
26. D'Onofrio, E.E.; Fiore, M.M.; Romero, S.I. Return periods of extreme water levels estimated for some vulnerable areas of Buenos Aires. *Cont. Shelf Res.* **1999**, *19*, 1681–1693. [CrossRef]
27. Li, K.; Li, G.S. Vulnerability assessment of storm surges in the coastal area of Guangdong Province. *Nat. Hazard.* **2013**, *68*, 1129–1139. [CrossRef]
28. Vicente-Serrano, S.M.; Beguería-Portugués, S. Estimating extreme dry-spell risk in the middle Ebro valley (northeastern Spain): a comparative analysis of partial duration series with a general Pareto distribution and annual maxima series with a Gumbel distribution. *Int. J. Clim.* **2003**, *23*, 1103–1118. [CrossRef]
29. China Oceanic Information Network (COIN). China Sea Level Bulletin 2017. Available online: http://www.nmdis.org.cn/hygb/zghpmgb/2017nzghpmgb/ (accessed on 10 September 2018).

30. IPCC. Summary for Policymakers. In *Climate Change 2013: The Physical Science Basis. Contribution of Working Group I to the Fifth Assessment Report of the Intergovernmental Panel on Climate Change*; Stocker, T.F., Qin, D., Plattner, G.-K., Tignor, M., Allen, S.K., Boschung, J., Nauels, A., Xia, Y., Bex, V., Midgley, P.M., Eds.; Cambridge University Press: Cambridge, UK; New York, NY, USA, 2013.
31. Nie, Y.; Tang, C.; Cheng, Z. Projections of Maximum Sea Level Recurrence Interval near Shenzhen at the End of the 21th Century. *Trop. Geogr.* **2016**, *36*, 901–905.

© 2020 by the authors. Licensee MDPI, Basel, Switzerland. This article is an open access article distributed under the terms and conditions of the Creative Commons Attribution (CC BY) license (http://creativecommons.org/licenses/by/4.0/).

Technical Note

Multi-Method Tracking of Monsoon Floods Using Sentinel-1 Imagery

Giuseppe Ruzza [1],*, Luigi Guerriero [1], Gerardo Grelle [2], Francesco Maria Guadagno [1] and Paola Revellino [1]

[1] Department of Science and Technology, University of Sannio, 82100 Benevento, Italy; luigi.guerriero@unisannio.it (L.G.); guadagno@unisannio.it (F.M.G.); paola.revellino@unisannio.it (P.R.)
[2] Department of Civil, Building and Environmental Engineering (DICEA), Sapienza University of Rome, 00184 Rome, Italy; gerardo.grelle@uniroma1.it
* Correspondence: giuseppe.ruzza@unisannio.it

Received: 24 September 2019; Accepted: 29 October 2019; Published: 31 October 2019

Abstract: Floods cause great losses in terms of human life and damages to settlements. Since the exposure is a proxy of the risk, it is essential to track flood evolution. The increasing availability of Synthetic Aperture Radar (SAR) imagery extends flood tracking capabilities because of its all-water and day/night acquisition. In this paper, in order to contribute to a better evaluation of the potential of Sentinel-1 SAR imagery to track floods, we analyzed a multi-pulse flood caused by a typhoon in the Camarines Sur Province of Philippines between the end of 2018 and the beginning of 2019. Multiple simple classification methods were used to track the spatial and temporal evolution of the flooded area. Our analysis indicates that Valley Emphasis based manual threshold identification, Otsu methodology, and K-Means Clustering have the potential to be used for tracking large and long-lasting floods, providing similar results. Because of its simplicity, the K-Means Clustering algorithm has the potential to be used in fully automated operational flood monitoring, also because of its good performance in terms of computation time.

Keywords: sentinel-1; SAR; flood; image classification; clustering; monsoon; Philippines

---

1. Introduction

Floods are among the most frequent and widespread natural hazards in the world. Being related to intense and/or extreme weather events, they cause great losses in terms of human life and damage to commercial and productive sites, infrastructures, and agriculture [1,2]. Particularly, it has been estimated that floods are responsible for approximately 40% of the total damage caused by natural hazards [3,4]. Exposure to flooding is considered a proxy of the risk [5], so that an evaluation of the extent of potentially inundated areas is crucial for hazard and risk assessment and represents the basis for land planning and policy decisions oriented towards flood mitigation (i.e., occupation restrictions, recommended uses, and flood insurance plan development). Flood hazard and risk evaluations can be completed using the results of deterministic hydrodynamic models that simulate water movement across the floodplain. In the presence of monitoring (fluvial stage or discharge) and topographic data, statistical models associated with GIS processing can provide a basis for such kinds of analyses [6–8]. Both deterministic and statistical models need to be calibrated and validated using available flood data in terms of spatial extent, persistence, and frequency.

Data derived from a number of satellite platforms can be used to image floods, providing a basis for a rapid and effective response to natural disasters. Satellite observations have the advantage of covering large areas with an increasingly short revisiting time, which makes them able to support continuous observation and operation monitoring of floods [9,10]. Among different satellite data, Synthetic Aperture Radar (SAR) products provide an opportunity to image floods because of their all-weather and day/night capability [11,12] and their sufficient resolution for urban and suburban mapping [13,14]. SAR observation capabilities through clouds allow tracking flood events connected to prolonged rainfall [15]. SAR data have been widely used to study this kind of event in different contexts and have the potential to support surface water operational monitoring [16]. Giustarini et al. [17] and Mason et al. [18] used SAR data for flood detection in an urban area. They provided a way to image floods also in contexts where it is difficult to separate water by land. For instance, Martinis et al. [19] used Sentinel-1 data to improve flood monitoring in arid areas, where the similarity between radar backscattering of open water and sand surfaces led to an overestimation of the water extent. SAR data allow floods to be imaged over very large areas. In this context, Xing et al. [20] monitored monthly changes in surface water of Dongting Lake. Space-born L-Band SAR data were used by Chapman et al. [21] to map regional inundation events on the continental scale. Water flood information produced by SAR imagery can be also associated with additional remote sensing data (e.g., multispectral and optical imagery) for detecting the flooding extent and evolution in what we could define as a multi-data/methods approach. For instance, Refice et al. [22] used multi-sensor and multi-temporal remote sensing approaches to characterize flooding that affected part of the Strymonas river basin, a transboundary river with its source in Bulgaria, which flows then through Greece up to the Aegean Sea. Hakdaoui et al. [23] used radar and optical data to extract geomorphological information after a flash flood event in a Saharan arid region, and Shuman et al. [24] used SAR data and aerial photography to track urban flood dynamics.

Many methodologies have been proposed in the literature for the identification of flooding from SAR images. Texture recognition algorithms [25], histogram thresholds [26], and various multi-temporal change detection methods [17] are examples of these methodologies. Bioresita et al. [27] constructed an automatic chain process for surface water extraction. An unsupervised method, based on stochastic subspace ensemble-learning, was proposed by the authors of [28]. Schlaffer et al. [29] used a harmonic analysis and change detection to extract flooded areas, and an unsupervised approach, based on a generalized Gaussian model, to automatically detect surface change was used by the authors of [30]. Many authors (e.g., [31–33]) proposed and used an automated method for the extraction of surface water from SAR data based on supervised and unsupervised approaches. Additional authors like Bayik et al. [34] used multiple classification methods to improve flood mapping from SAR images. Since such methods have a variable degree of complexity (i.e., required parameters in relation to algorithm complexity) and attitudes from an automatic operational monitoring perspective, a number of comparisons have been proposed in order to underline the advantages and disadvantages of each method (e.g., [35]).

On this basis, and in order to further contribute to a better evaluation of the potential of SAR imagery in tracking large and long-lasting multi-pulse floods, we comparatively analyzed simple classification methods that might be suitable for automatic flood monitoring. Notably, the aim of this paper was to demonstrate the suitability of SAR products to image floods and quantify the magnitude of variation in the flooding extent derived by the use of multiple classification methods for the same event. In this perspective, we used Sentinel-1 imagery to track the evolution of a flood event that occurred in the monsoon area of the Philippines. The event, caused by torrential rains, occurred at the end of December 2018, persisted until the end of February 2019, and affected approximately 680,000 people. The area of Camarines Sur Province, in the eastern Philippines, suffered the most substantial effects. This event represents a very important case in history because of its effects, the extent of the involved area, and its duration, which makes it suitable for a SAR-based analysis and comparison between multiple classification methods.

2. Study Area

This study area included the Camarines Sur Province and a small portion of the Albay Province (Figure 1) located in the Bicol Region in Luzon of the Philippines. The Camarines Sur Province occupies the central section of the Bicol Peninsula and is the largest province in the Bicol Region. The Bicol River is the main watercourse of the province and is surrounded by Mount Isarog (1966 m asl) and Mount Iriga (1196 m asl). The eastern portion of the province lies on the mountainous peninsula of Caramoan, which faces the island of Catanduanes to the east. The Bicol River drains the central and southern parts of the province toward the San Miguel Bay. Mount Iriga is surrounded by three lakes named Buhi, Bato, and Baao. The Albay Province is generally mountainous with scattered valleys. On the eastern part of the province is a line of volcanic mountains starting with the northernmost Malinao in Tiwi and followed by Mount Masaraga (1328 m asl) and the Mayon Volcano (2463 m asl).

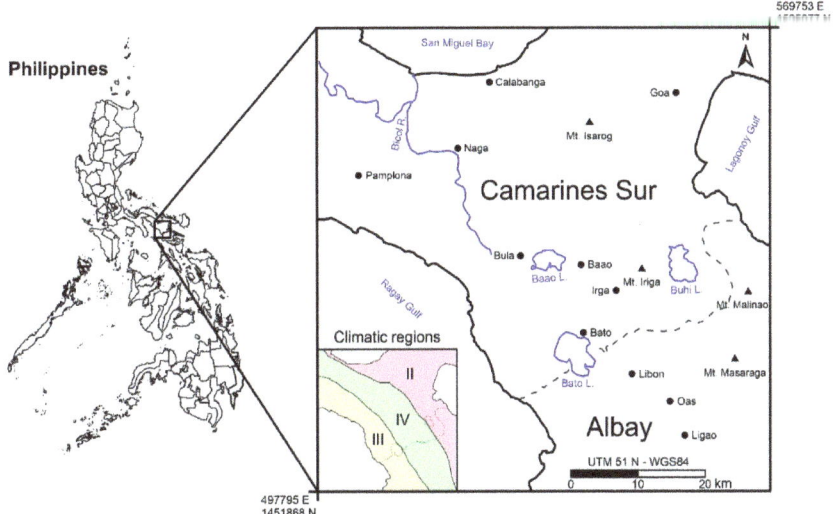

Figure 1. Map showing the study area (dashed line divides Camarines Sur and Albay provinces). The blue line indicates the major river, lakes are represented by blue polygons, towns are reported using black dots, and major reliefs are depicted by black triangles. The inset map shows the climatic regions of the study area.

The Philippines archipelago has a tropical and strongly monsoonal climate characterized by relatively high temperature, oppressive humidity, and plenty of rainfall. Two main seasons alternate for the majority of islands, a wet season dominated by monsoon rainfall and a dry season. Some areas experience rainfall throughout the year, and clear season alternations are absent. Four climate regions are recognizable across the Philippines archipelago characterized by different distributions of rainfall [36,37]. The study area spans across three of these four climatic regions (Figure 1): region II is characterized by the absence of a dry season with a very pronounced maximum rainfall during the months of November and December, region III is characterized by a slightly drier seasons between November and April and a wet season between May and October, and region IV is characterized by rainfall distributed throughout the entire year. From June to December, tropical cyclones (typhoons) often strike the Philippines. Most of these storms come from the southeast and are heaviest in Samar, Leyte, south-central Luzon, and the Batan Islands, and, when accompanied by floods or high winds, they may cause great loss of life and properties.

In December 2018, as reported by the Department of Science and Technology (PAGASA) of the Republic of Philippines (http://bagong.pagasa.dost.gov.ph/tropical-cyclone), the study area was flooded as a consequence of a typhoon induced by a tropical depression (named "*Usman*"). This was the 21st and the last cyclone that intersected the Philippines in 2018. The "*Usman*" depression, developed from tropical disturbances, resulted in widespread heavy rainfall over large portions of Southern Luzon and Eastern Visayas. From 28 to 29 December 2018, prolonged and excessive rainfall over these areas resulted in multiple landslides and floods. Of the 139 stations with reliable data, the highest 2-day rainfall was estimated at 573.2 mm and was recorded in Daet (Camarines Norte), 50 km away from the study area. In the week preceding the event, rainfall accumulations in excess of 100 mm were already observed by multiple rain gauges over the eastern section of Luzon and Visayas, especially over Bicol Region mainland, Samar island, Aurora, Quezon, and Rizal. The event caused hundreds of casualties and millions of dollars in damages to personal and public property. According to the report of the Department of Social Welfare and Development (DSWD), more than 680,000 people were affected by the storm. Nearly 55% were located primarily in the province of Camarines Sur in the Bicol Region. Over 1900 homes were destroyed, and more than 15,000 were damaged. Agricultural losses were estimated at PHP 2 billion (US $37 million), with over 56,000 farmers and fishers affected, according to the United Nations Food and Agriculture Organization.

3. Data and Methods

3.1. Data

To quantify the extent of the flooded area, i.e., the area covered by open water, and track its spatial and temporal evolution, we used multiple Sentinel-1 A/B SAR images. The satellite data were downloaded from the Copernicus Open Data Access Hub (https://scihub.copernicus.eu/dhus/#/home). Sentinel-1 data products consist of imagery with a medium resolution (10–20 m) that represents the radar echo from the surface of the Earth of the signal emitted by the onboard antenna of the satellite. The Sentinel-1 satellite constellation is formed by two satellites (S-1 A and S-1 B) characterized by an acquisition frequency of six days. Each satellite is equipped with a 5.405 GHz C-band ($\lambda \approx 5.6$ cm) imager payload (CSAR). The CSAR instrument supports operation in dual-polarization (HH + HV and VV + VH). The Sentinel-1 satellites can operate in three imaging modes for various observation approaches, spatial resolutions, and swath widths [38]. In this study, we used a total of 24 images with VV polarization. We chose this polarization because it characterized all of the selected data; only a limited number of images were acquired in double polarization VV + HH. VV polarization, in some cases, provides better results in detecting open water features [39,40]. The selected images had the same angle of incidence (38.8°) and footprint and were acquired in descending mode. The selected images covered the time period between 13 November 2018, prior to the typhoon, and 18 March 2019, after the typhoon. The first selected image was acquired 78 days before the main flooding event, occurring between 28 and 29 December 2018. This long pre-event time span allowed us to observe the normal fluctuations of the open water area before the flood.

3.2. Data Pre-Processing

Waterbody identification and extraction were completed considering the difference in the backscattering coefficient (Gamma $\gamma°$) between open water and the land surface. This difference is related to the roughness of the target surface, which is regulated by the Rayleigh criterion. Since the open water theoretically has no roughness, the incident radiation is reflected away from the sensor; thus, the backscattering coefficient related to its flat surface is lower than the land surface [41–43]. In contrast, the land surface that has a variable roughness has a variable backscattering coefficient typically higher than that characterizing open water. SAR imagery, downloaded by Sentinel Data Hub, were pre-processed using the Sentinel Application Platform (SNAP) toolkit distributed by ESA (https://step.esa.int/main/toolboxes/snap/). The products were first cropped across the area of interest to allow for shorter processing times. Subsequently, radiometric calibration was applied, and terrain was corrected to the Gamma band (i.e., backscatter coefficient $\gamma°$) [44]. To complete this last step, the STRM 1 arc-second digital elevation model (DEM) was used. Finally, in order to obtain high-quality SAR images and reduce the noise, we applied the Lee-Sigma filter [45,46] characterized by a combination of 5×5 and 7×7 windows.

As the last step of image pre-processing, calibrated and filtered SAR images were cropped over the land of the study area. This was done in order to remove (from images and relative histograms) the changing backscattering effects induced at the ocean surface by the difference in wind conditions that generating waves mimicked the surface roughness, which produced a double-bounce effect [47]. After pre-processing, the rearranged backscatter intensities were converted into dB (decibels). Figures 2 and 3 show a selection of 12 SAR images, out of 24 used, in which it is possible to visually identify the different flood traces of the events.

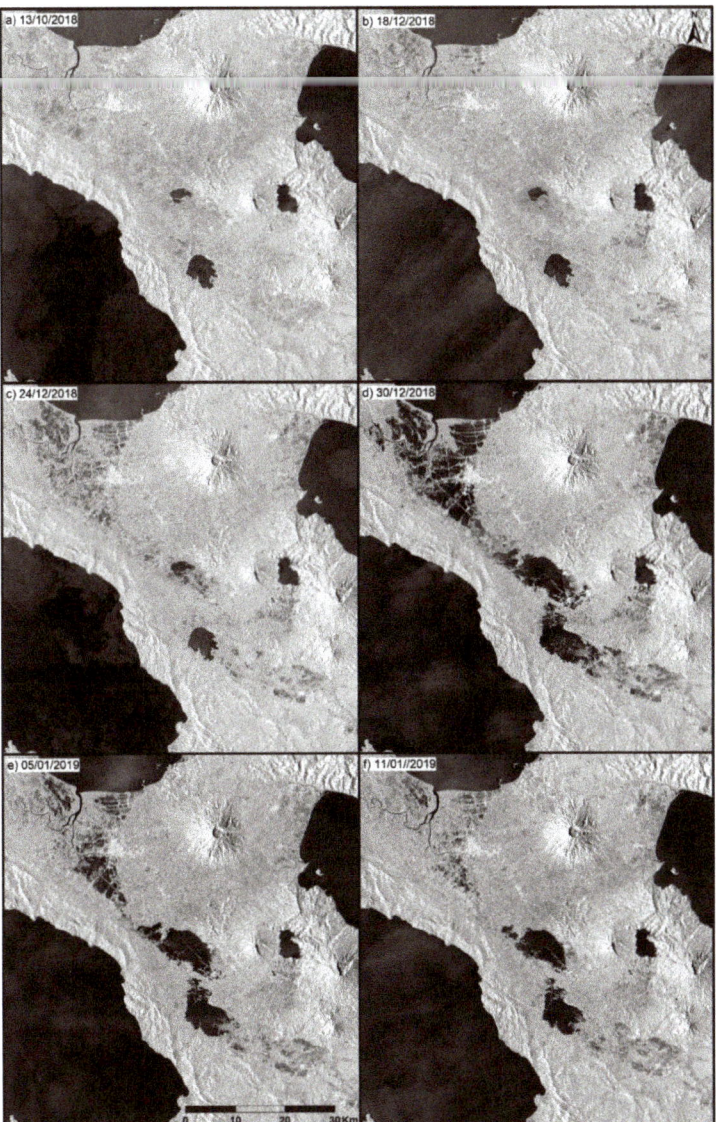

Figure 2. Examples of pre-processed Synthetic Aperture Radar (SAR) images of the study area in the period between 13 October 2018, and 11 January 2019. The dark areas represent open water features and shadow zones.

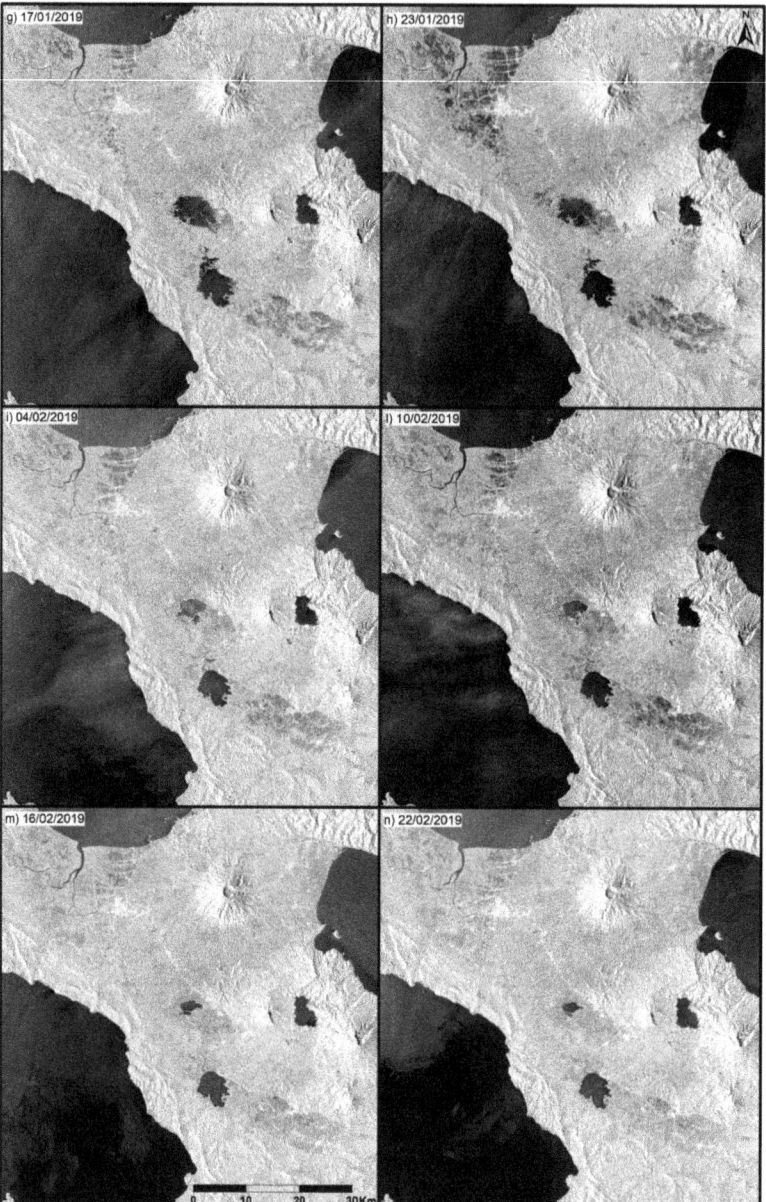

Figure 3. Examples of pre-processed SAR images of the study area in the period between 17 January 2019, and 22 February 2019. The dark areas represent open water features and shadow zones.

3.3. Extraction of the Water Body Area

Figure 4 summarizes the workflow used to process images and quantify the total water area in the reference time period. In order to identify the open water features to determine the total area covered by water and its spatial and temporal evolution due to monsoon flood events, three different classification methods were applied that might have the potential to be used in operational monitoring.

Our multi-method approach allowed us to compare specific results and quantify relative differences (e.g., [48,49]) in order to identify the best method to be used from an operational monitoring perspective. Particularly, we used two methods based on threshold identification and a third based on unsupervised K-Means Cluster analysis. Threshold-based methods are widely used because of their efficiency in identifying flooded areas in SAR images [50–53]. These methods consisted of defining a fixed value for the backscattering coefficient (i.e., intensity) that split the histogram in two subsets (clusters), one subset representing the land surface and the other representing the open water. Pixels with a low grayscale intensity typically corresponded to water bodies, while pixels with high values corresponded to the land (in our case, background). There are several techniques (supervised and unsupervised) to find the optimal threshold value [54–57]. Among these, for our comparative analysis, we selected visual identification of the threshold guided by the Valley Emphasis criterion, the Otsu method, and the unsupervised classification method based on K-Means Clustering.

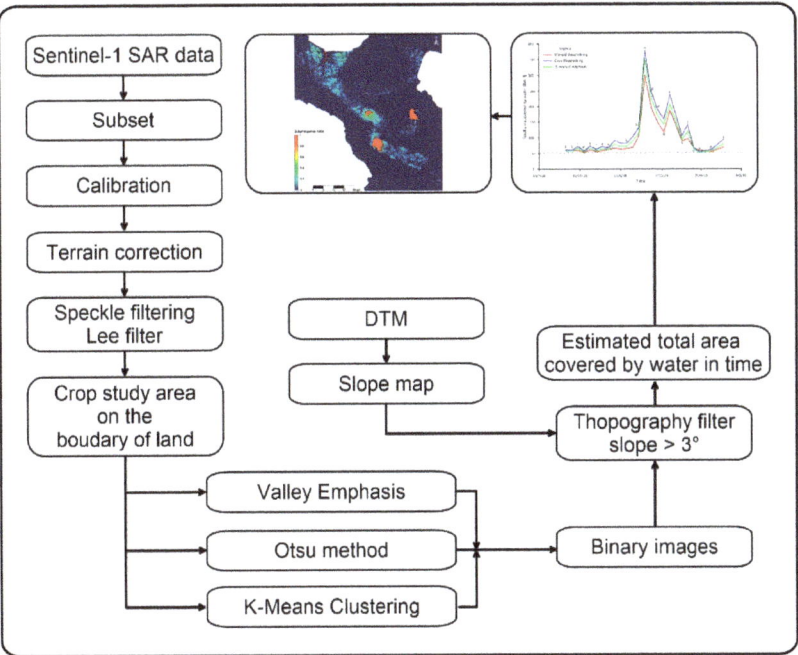

Figure 4. Processing workflow of SAR imagery for estimating the total surface covered by water across the study area in the reference period.

The first method consisted of visually inspecting a grayscale histogram of each SAR image and applying the Valley Emphasis criterion [58–60]. For the application of this criterion, we first analyzed the grayscale histogram in SNAP and, subsequently, chose the minimum backscatter intensity value that corresponded to the valley between the two peaks, in the case of bimodal distribution, or to the bottom rim of a single peak for a unimodal distribution. Once identified, we converted this intensity value from dB to a "grayscale" value between 0 to 255 (8-bit images). The identified threshold value was then used as a basis for manual image classification. Image classification was consistently completed in the GIS environment (QGIS 3.7), assigning 0 to pixels with values higher than the threshold (i.e., land) and 1 to pixels with values lower than the threshold (i.e., water). After classifying the images, and considering that low backscattering intensity values can be also related to the presence of shadow zones (i.e., not only to the presence of water), especially in presence of complex topography, we used a further classification method based on topographic attributes to identify and exclude these areas from our analysis. Notably, we generated a slope map using the available STRM 1 arc-second DEM and used a slope threshold of 3° to exclude the low backscattering areas with slopes higher than the threshold.

The second method used to find the correct threshold value was the Otsu method. This method is widely used in applications where it is necessary to split grayscale images into two different classes. This method automatically chooses the threshold value from the grayscale histogram on the basis of the minimum within-class variance or the maximum between-class variance [61]. The Otsu method is commonly used because of its simplicity, and the best results are obtained if the image histogram is characterized by a bimodal or multimodal distribution. In the presence of a unimodal distribution, this method fails. The Otsu method for finding an optimal threshold value for image classification was applied in the Matlab™ environment. The Otsu threshold was then used to classify the images with the same procedure used as the previous method.

The last method was the K-Means Cluster analysis. This clustering method is a statistical technique widely used for dimensional reduction and has the potential to be used for unsupervised analysis of SAR images [62,63]. In our case, K-Means Clustering was used for dividing the image in two classes (i.e., water and land clusters). This method groups pixels on the basis of their grayscale level distribution and standard deviation. The purpose of this algorithm is to reduce variability within clusters, and the objective function is the sum of square distances between cluster centers and its assigned pixel value [64]. This clustering method, implemented in the SNAP toolbox, was applied considering two clusters, 30 interactions, and 32,000 random seeds. After cluster identification, the topographic filter and subsequent manual classification of the images were completed. Finally, we estimated the total water area in each image.

Once the spatial extent of open water features across the study area was evaluated, to further analyze the spatial and temporal distribution of flood intensity and persistence, and its change between the different methods, we estimated the submergence ratio across the whole of the study area and the reference period, and we compared this ratio in terms of arithmetic differences between products derived by our data (i.e., water area extent) obtained by different methods. A submergence ratio equal to 1 indicated the constant presence of water in the reference area, while a value of 0 indicated the constant absence of water in the reference area and across the reference period. A pixel value between 1 and 0 indicated that the reference area was only temporally submerged. This condition is consistent with the occurrence of the flood event, and the difference in the submergence ratio is representative of flood persistence and might be related to the water level. The submergence ratio map was constructed on the basis of the extent of open water features derived using the K-Means Clustering classification method. To determine the variation in the submergence ratio derived by the use of open water coverage derived by the three classification methods, we calculated the submerged ratio difference by iterative subtraction.

4. Results and Discussion

Figures 5 and 6 show several examples of results in terms of open water identification from the different methods used to select the optimal threshold value of backscattering intensity. The first example reported in Figure 5a,c,e represents the water coverage at the beginning of the monitoring period before the occurrence of the rainfall events responsible for the flooding. The second example reported in Figure 5b,d,f represents the water coverage immediately after the first flood pulse induced by rainfall. Figure 6a,c,e represent the water coverage immediately after the second flood pulse induced by rainfall. The second example reported in Figure 6b,d,f represents the water coverage immediately after the third flood pulse induced by rainfall. For comparison purposes, in each map of Figures 5 and 6, polygons representing water bodies as depicted in the SWBD dataset (red polygon, e.g., [65]) were overlaid with areas classified as water bodies (blue area). The SWBD is the worldwide open water bodies boundary in vector format, which were generated by the National Geospatial Intelligence Agency [66] and represent the maximum extent of open water features. A comparison of the results in terms of water coverage indicated the consistency of all of the methods across most of the study area, with local relative underestimation of the Valley Emphasis method. An example is in the northeastern sector of the study area. No differences between the Otsu and K-Means Clustering were appreciable at the scale of Figures 5 and 6.

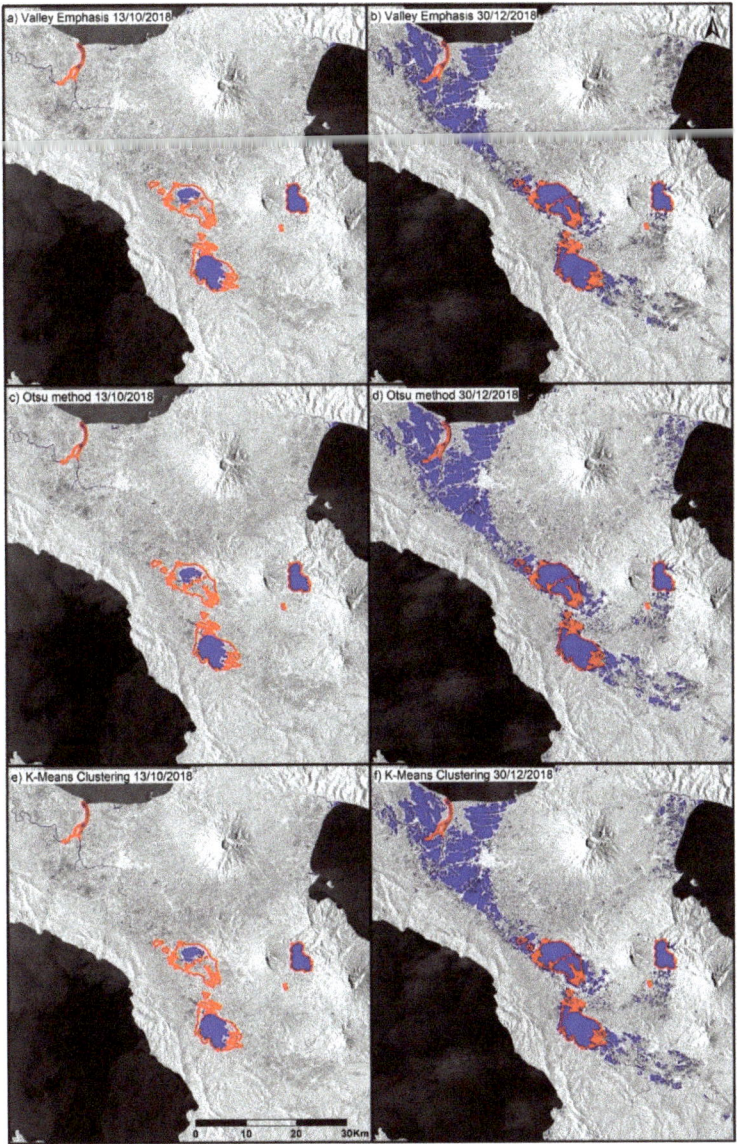

Figure 5. Examples of results in terms of open water extent obtained by multiple methods. Blue areas represent the extracted open water features, and red polygons represent SWBD water body boundaries. (**a,c,e**) represent the extent of open water features before the rainfall events responsible for the flooding. (**b,d,f**) represent the extent of open water features after the first flood pulse (main).

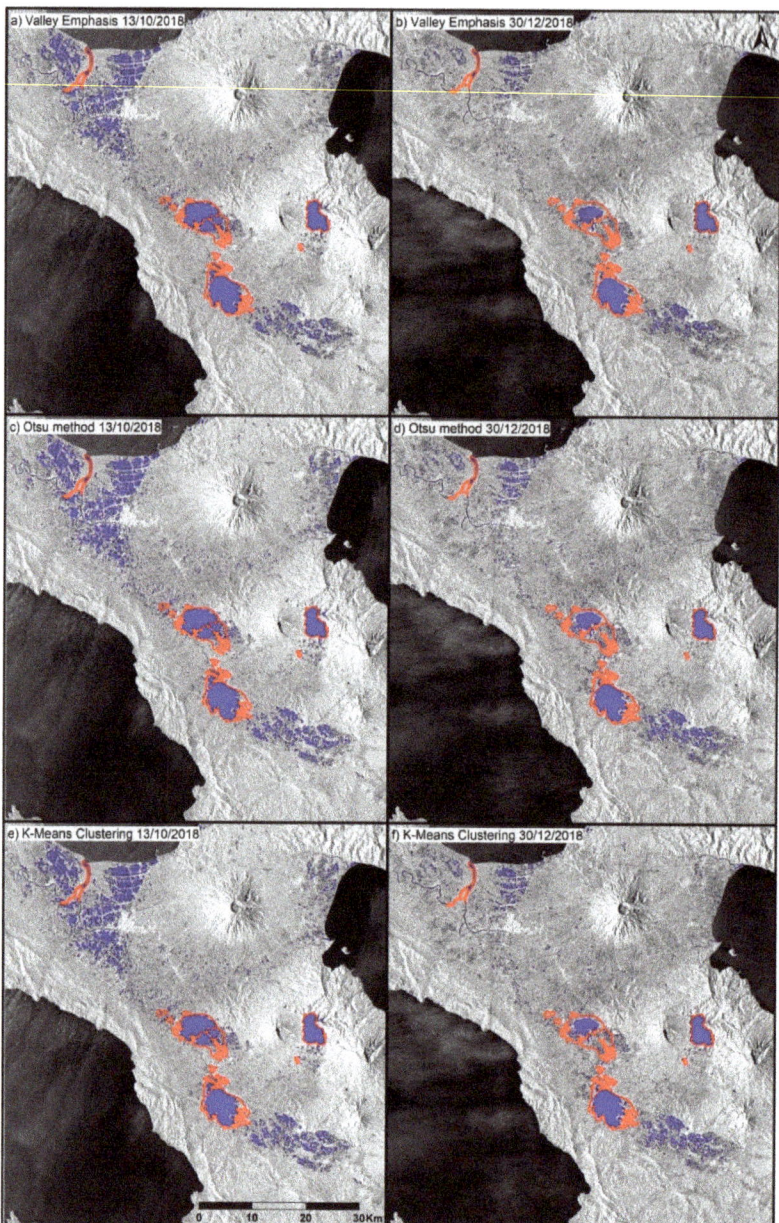

Figure 6. Examples of results in terms of open water extent obtained by multiple methods. Blue areas represent the extracted open water features, and red polygons represent SWBD water body boundaries. (**a**,**c**,**e**) represent the extent of open water features after the second rainfall event responsible for the flooding. (**b**,**d**,**f**) represent the extent of open water features after the third flood pulse.

Figure 7 shows the evolution of the open water features during the period of interest, including the rainfall events responsible for multiple flood pulses, in terms of total extent for each considered method. Particularly, the red, blue, and green lines represent the total flooded area extracted by Valley

Emphasis, Otsu, and K-Means Clustering methods, respectively. All three methods showed the same pattern of water coverage over time with some variable differences in terms of total estimation. Before the first flood pulse, from 13 October to 24 December the graph indicated a slow increase of water coverage, from a minimum coverage area of about 52 km^2 (grey dashed line in the graph). After this first period, three individual flood pulses with decreasing intensities were identified in terms of total water area increase by our multi-method analysis. According to the PAGASA report, the first main flood pulse occurred between 30 December 2018 and 1 January 2019. A second and a third minor flood pulse followed the first occurring between 17 January 2019, and 16 February 2019. After these water surface increases, the progressive recession of open water features induced a consistent decrease of the estimated total extent of the water surface that, between 22 February 2019, and 6 March 2019, matched the extent of water features before the flood.

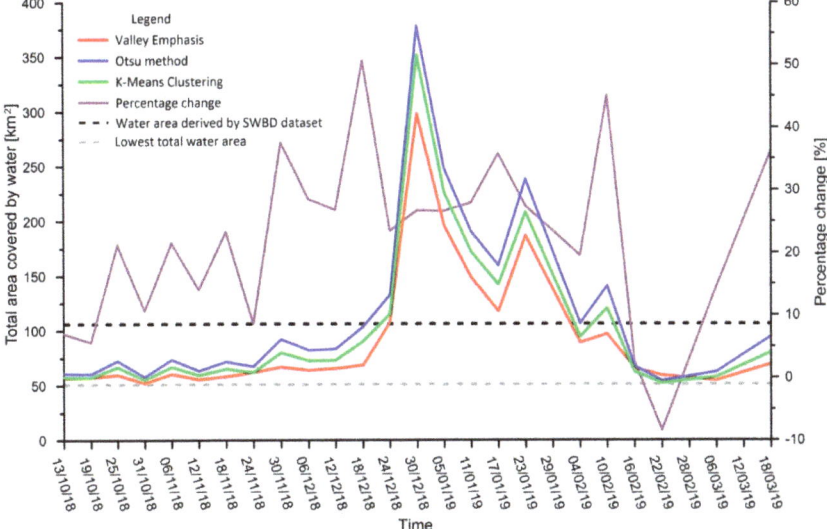

Figure 7. Graph showing the evolution of open water feature surface coverage estimated with the Valley Emphasis, Otsu, and K-Means Clustering methods (i.e., red, blue and green lines, respectively) in the study area over the monitoring period. The purple line shows the percentage difference of water area between the Valley Emphasis and Otsu methods, which show the lowest and the highest estimation value. The gray line represents the minimum value of the water area estimated by our analysis before the occurrence of the flood, and the black line represents the water area derived by the SWBD dataset.

Even if the three methods showed consistent trends in time, a number of differences in the total estimated water area with the different methods are depicted in the graph of Figure 7. The purple line of the graph provides an overview of this change in terms of the maximum percentage change. Indeed, the average difference of approximately 20% was observable. Because of the different estimations of the open water area at the beginning of the period of interest and the absence of an official estimation, our analysis was able to provide only the relative change of the open water area coverage estimated for each different flood pulse occurring in the reference period. Notably, our analysis indicated that the first (main) flood pulse produced an increase in total water area between 245 and 326 km^2, the second pulse was responsible for an increase between 135 and 186 km^2, and the third event induced an increase between 45 and 88 km^2. As described above and observable from the graph of Figure 7, the minimum value of open water area coverage was consistently derived by the Valley Emphasis method, while the maximum value was consistently derived by the Otsu method. The K-Means Clustering method showed water coverage values between the minimum (i.e., Valley Emphasis method) and the maximum (i.e., Otsu method). The flood-induced increases in the open water area estimated with K-Means Clustering were equal to 300, 156, and 68 km^2 for the first, second, and third flood pulses, respectively. The shift from the other two methods ranged, in relative terms and considering the entire reference period, between 4% and 31%.

The Otsu method consistently showed higher open water surface values across the reference period. The tendency to overestimate water coverage by the Otsu classification method in comparison with the Valley Emphasis method was observed also by Ba Duy [59], who confirmed that this method outperformed the Otsu method in terms of absolute identification capabilities. In addition, the author of the study underlines the limitation of the Valley Emphasis method for extracting water by SAR images, including inefficient identification of mixed water pixels, confusion of water bodies with background noise, and difficulty in choosing an optimal threshold value for very large areas. Bin Cui et al. [28] indicated that conversely from threshold-based methods like the Otsu and Valley Emphasis, the unsupervised methods, like the K-Means Clustering, have great potential in the context of image classification.

The good performance of the unsupervised method in comparison with threshold-based methods can be related to its ability in capturing the spatial variability of the backscattering coefficient that represents a basis for a self-adaptation to different environmental conditions captured by SAR images. Our application confirmed that despite the lack of supervision and parameter estimation (e.g., threshold), the K-Means Clustering performed at least as good as the other methods, and possibly better than the Otsu. In this way, our interpretation is that the unsupervised methods, like the K-Means Clustering, have the potential to support SAR imagery-based flood mapping, also in long-term operational frameworks. This is also related to better results in terms of computational time [28]. It is interesting to mention that additional methods for water body classification, like Bimodal Histogram and Local Adaptive Thresholding [67], have been introduced and have the potential to support this kind of analysis. However, their applicability and reliability need to be further analyzed.

Figure 8a shows the persistence of the water in the reference period across the whole study area. In the submergence map of Figure 8a, permanent water bodies like the major lakes and the river are easily recognizable, as they are marked by a color corresponding to a value of the submergence ratio equal to 1. The Baao and Bato lakes expanded due to the flood events until they reached the boundary indicated in the SWBD dataset for a limited period of time (i.e., 20% of the reference time period). Most of the area hit by the flood stayed submerged between 5% and 25% of the time, with a number of spots in which the water persisted for approximately 40% of the reference time period. These spots were in the areas surrounding major lakes and next to the northern coastline of the study area. We expect that the temporal persistence of the flood was directly related to the water level and the presence of a draining channel.

Figure 8b–d, report the difference in the submergence ratios calculated with products derived by image classification and the selected methods. Notably, the most important differences were between submergence ratios calculated with the products derived by K-Means Clustering and Valley Emphasis based threshold definitions and between Otsu and Valley Emphasis based threshold definitions. This is consistent with the results of our analysis summarized in Figure 7.

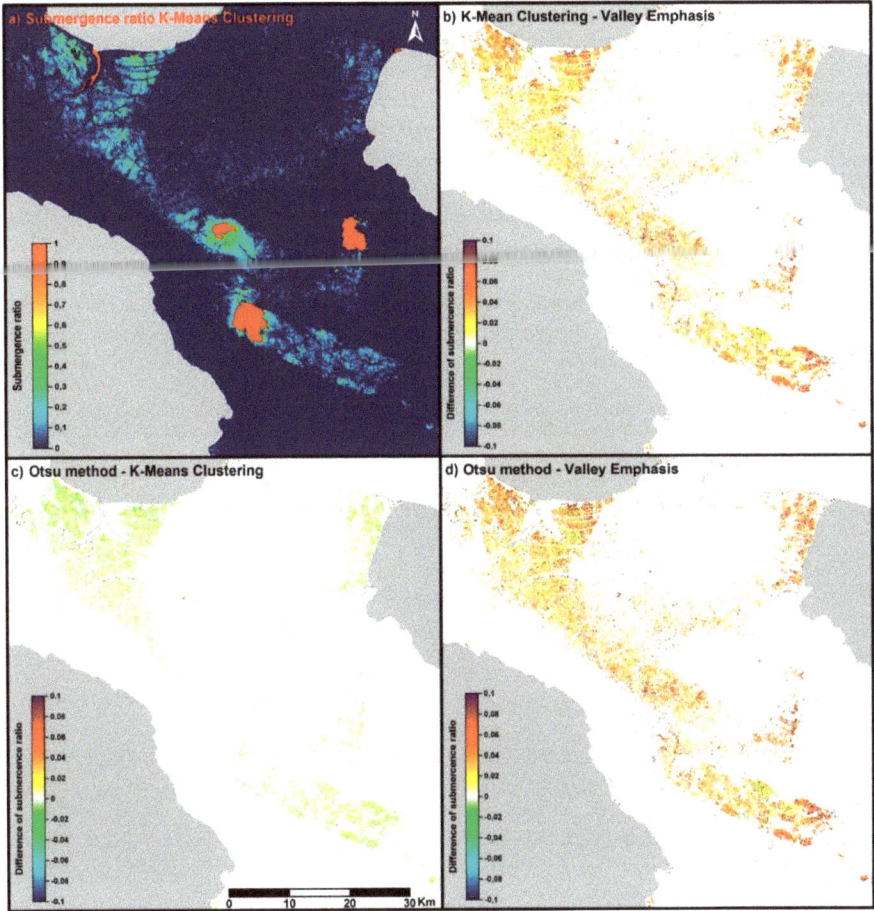

Figure 8. The map of (**a**) shows the submergence ratio (i.e., the flood persistence) in the reference period. A pixel value equal to 1 corresponds to an always flooded area. A value of 0 corresponds to a never flooded area. The maps of (**b–d**) are the differences in submergence ratios calculated using open water features extracted by different methods.

5. Conclusions

Our analysis indicates that Sentinel-1 data have the potential to be used in flood tracking and for monsoon events characterized by long-term rainfall associated with persistent cloud coverage, which excludes the possibility of using the other types of satellite products, like optical imagery. In this study, the multi-pulse flood caused by the *Usman* tropical depression in the Camarines Sur Province of the Philippines was analyzed with the aim to provide quantitative flooding data and test multiple methods from a relative perspective connected to the absence of accessible official data. All of the pulses characterizing the flood event occurred between 24 December 2018, and 23 January 2019, remembered for its dramatic effects over the Camarines Sur Province, and they were detected and characterized in terms of the open total flooded area. Since the discrimination between water features and the land surface might be a very challenging task in terms of both accuracy and computation time, in our study, the open water features extraction was completed with three different methods: the Valley Emphasis based manual threshold identification, the Otsu method, and the K-Means Clustering method. All of the methods identified a shape consistent with the flooded area, with slight differences in dimensions (i.e., total area) observed over the monitoring period. These differences seem to be not correlated to the size of the total area covered by open water. Although an official estimation of the flooded area is not available, these methods provide a total flooded area that approximately averaged the results obtained with the Valley Emphasis based manual threshold identification and the Otsu method. The K-Means Clustering method is advantageous because (i) it is unsupervised, (ii) it is fully automatic, and (iii) it returned better results in terms of the computation time. A fast-computational time is a great advantage when it is necessary to have a quick overview of a flood situation. On this basis, this method might have the potential to be used in an operational flood monitoring perspective. In this framework, SAR images are an important means to study flood phenomena because of their day and night operational capabilities and their increasing frequency of acquisition related to the number of satellite constellations. In addition, these capabilities are enhanced by the free availability of data, like those derived by Sentinel-1 satellites.

Author Contributions: Conceptualization, G.R. and L.G.; methodology, G.R.; software, G.R.; validation, G.R., L.G. and G.G.; formal analysis, L.G.; investigation, G.R.; resources, P.R.; data curation, G.R. and F.M.G.; writing—original draft preparation, G.R. and L.G.; writing—review and editing, L.G., G.G., P.R. and F.M.G.; visualization, G.R.; supervision, P.R. and F.M.G.; project administration, P.R.; funding acquisition, P.R.

Funding: This research was supported by "Research University Funds" of the University of Sannio, Resp.: Paola Revellino.

Acknowledgments: We thank to two anonymous reviewers for their constructive comments.

Conflicts of Interest: The authors declare no conflict of interest.

References

1. Revellino, P.; Guerriero, L.; Mascellaro, N.; Fiorillo, F.; Grelle, G.; Ruzza, G.; Guadagno, F.M. Multiple Effects of Intense Meteorological Events in the Benevento Province, Southern Italy. *Water* **2019**, *11*, 1560. [CrossRef]
2. Paul, S.H.; Sharif, H.O. Analysis of Damage Caused by Hydrometeorological Disasters in Texas, 1960–2016. *Geosciences* **2018**, *8*, 384. [CrossRef]
3. Ologunorisa, T.E.; Abawua, M.J. Flood risk assessment: A review. *J. Appl. Sci. Eniron. Manag.* **2005**, *9*, 57–63.
4. Re, M. Natural Catastrophes 2015, Annual Figures. Munich Re NatCat Service. 2016. Available online: https://www.munichre.com/site/corporate/get/params_E1254966961_Dattachment/1130647/Munich-Re-Overview-Natural-catastrophes-2015.pdf (accessed on 9 October 2019).
5. Jongman, B.; Ward, P.J.; Aerts, J.C.J.H. Global exposure to river and coastal flooding: Long term trends and changes. *Glob. Environ. Chang.* **2012**, *22*, 823–835. [CrossRef]
6. Di Baldassarre, G.; Schumann, G.; Bates, P.D.; Freer, J.E.; Beven, K.J. Flood-plain mapping: A critical discussion of deterministic and probabilistic approaches. *Hydrol. Sci. J.* **2010**, *55*, 364–376. [CrossRef]
7. Alfonso, L.; Mukolwem, M.M.; Di Baldassarre, G. Probabilistic flood maps to support decision-making: Mapping the value of information. *Water Resour. Res.* **2016**, *52*, 1026–1043. [CrossRef]
8. Guerriero, L.; Focareta, M.; Fusco, G.; Rabuano, R.; Guadagno, F.M.; Revellino, P. Flood hazard of major river segments, Benevento Province, Southern Italy. *J. Maps* **2018**, *14*, 597–606. [CrossRef]
9. Hoet, P.H.M.; Geys, J.; Nemmar, A.; Nemery, B. *NATO Science for Peace and Security Series C, Environmental Security*; Korgan, F., Powell, A., Fedorov, O., Eds.; Springer: New York, NY, USA, 2011.
10. Klemas, V. Remote Sensing of Floods and Flood-Prone Areas: An Overview. *J. Coast. Res.* **2015**, *31*, 1005–1013. [CrossRef]
11. Matgen, P.; Schumann, G.; Henry, J.B.; Hoffmann, L.; Pfister, L. Integration of SAR-derived river inundation areas, high-precision topographic data and a river flow model toward near real-time flood management. *Int. J. Appl. Earth Obs. Geoinf.* **2007**, *9*, 247–263. [CrossRef]
12. Kussul, N.; Shelestov, A.; Shakun, S. Intelligent computations for flood monitoring. In Proceedings of the XIVth International Conference 'Knowledge-Dialogue-Solution' KDS, Varna, Bulgaria, 23 June–3 July 2008.
13. Shen, X.; Wang, D.; Mao, K.; Anagnostu, E.; Hong, Y. Inundation Extent Mapping by Synthetic Aperture Radar a Review. *Remote Sens.* **2019**, *11*, 879. [CrossRef]

14. Ban, Y.; Jacob, A.; Gamba, P. Spaceborne SAR data for global urban mapping at 30 m resolution using a robust urban extractor. *ISPRS J. Photogramm. Remote Sens.* **2015**, *103*, 28–37. [CrossRef]
15. Kiage, L.M.; Walker, N.D.; Balasubramanian, S.; Baras, J. Application of Radarsat-1 synthetic aperture radar imagery to assess hurricane-related flooding of coastal Louisiana. *Int. J. Remote Sens.* **2005**, *26*, 5359–5380. [CrossRef]
16. Bolanos, S.; Stiff, D.; Brisco, B.; Pietroniro, A. Operational Surface Water Detection and Monitoring Using Radarsat 2. *Remote Sens.* **2016**, *8*, 285. [CrossRef]
17. Giustarini, L.; Hostache, R.; Matgen, P.; Schumann, J.-P. A Change Detection Approach to Flood Mapping in Urban Areas Using TerraSAR-X. *IEEE Trans. Geosci. Remote Sens.* **2013**, *51*, 2417–2430. [CrossRef]
18. Mason, D.C.; Davenport, I.J.; Neal, J.C.; Shumann, G.J.-P.; Bates, P.D. Near real-time flood detection in urban and rural areas using high resolution synthetic aperture radar images. *IEEE Trans. Geosci. Remote Sens.* **2012**, *50*, 3041–3052. [CrossRef]
19. Martinis, S.; Plank, S.; Cwik, K. The Use of Sentinel-1 Time-Series Data to Improve Flood Monitoring in Arid Areas. *Remote Sens.* **2018**, *10*, 583. [CrossRef]
20. Xing, L.; Tang, X.; Wang, H.; Fan, W.; Wang, G. Monitoring monthly surface water dynamics of Dongting using Sentinel-1 data at 10 m. *PeerJ* **2018**, *6*, e4992. [CrossRef]
21. Chapman, B.; McDonald, K.; Shimada, M.; Rosenqvist, A.; Schroeder, R.; Hess, L. Mapping Regional Inundation with Spaceborn L-Band SAR. *Remote Sens.* **2015**, *7*, 5440–5470. [CrossRef]
22. Refice, A.; D'Addabbo, A.; Lovergine, F.P.; Tijani, K.; Morea, A.; Nutricato, R.; Bovenga, F.; Nitti, D.O. Monitoring Flood Extent and Area Through Multisensor, Multi-temporal Remote Sensing: The Strymonas (Greece) River Flood. In *Flood Monitoring through Remote Sensing*; Springer Remote Sensing/Photogrammetry; Springer: Cham, Switzerland, 2018. [CrossRef]
23. Hakdaoui, S.; Emran, A.; Pradhan, B.; Lee, C.-W.; Fils, S.C.N. A Collaborative Change Detection Approach on Multi-Sensor Spatial Imagery for Desert Wetland Monitoring after a Flash Flood in Southern Morocco. *Remote Sens.* **2019**, *11*, 1042. [CrossRef]
24. Schumann, G.J.-P.; Neal, J.C.; Mason, D.C.; Bates, P.D. The accuracy of sequential aerial photography and SAR data for observing urban flood dynamics, a case study of the UK summer 2007 floods. *Remote Sens. Environ.* **2011**, *115*, 2536–2546. [CrossRef]
25. Schumann, G.; Henry, J.B.; Hoffmann, L.; Pfister, L.; Pappenberger, F.; Matgen, P. Demonstrating the high potential of remote sensing in hydraulic modelling and flood risk management. In Proceedings of the Annual Conference of the Remote Sensing and Photogrammetry Society with the NERC Earth Observation Conference, Portsmouth, UK, 6–9 September 2005.
26. Psomiadis, E. Flash flood area mapping utilizing Sentinel-1 radar data. In Proceedings of the SPIE Earth Resources and Environmental Remote Sensing/GIS Applications VII, Edinburgh, UK, 26–29 September 2016; p. 100051G.
27. Bioresita, F.; Puissant, A.; Stumpf, A.; Malet, J.-P. A Method for Automatic and Rapid Mapping of Water Surface from Sentinel-1 Imagery. *Remote Sens.* **2018**, *10*, 217. [CrossRef]
28. Cui, B.; Zhang, Y.; Yan, L.; Wei, J.; Wu, H. An Unsupervised SAR Change Detection Method Based on Stochastic Subspace Ensemble Learning. *Remote Sens.* **2019**, *11*, 1314. [CrossRef]
29. Schlaffer, S.; Matgen, P.; Hollaus, M.; Wagner, W. Flood detection from multi-temporal SAR data using harmonic analysis and change detection. *Int. J. Appl. Earth Obs. Geoinf.* **2015**, *38*, 15–24. [CrossRef]
30. Bazi, Y.; Bruzzone, L.; Melgani, F. An Unsupervised Approach Based on the Generalized Gaussian Model to Automatic Change Detection in Multiple SAR Images. *IEEE Trans. Geosci. Remote Sens.* **2005**, *43*, 874–887. [CrossRef]
31. Huang, W.; De Vries, B.; Huang, C.; Lang, M.W.; Jones, J.W.; Creed, I.F.; Carroll, M.L. Automated Extraction of Surface Water Extent from Sentinel-1 Data. *Remote Sens.* **2018**, *10*, 797. [CrossRef]
32. Benoudjit, A.; Guida, R. A Novel Automated Mapping of the Flood Extent on SAR Images Using a Supervised Classifier. *Remote Sens.* **2019**, *11*, 779. [CrossRef]
33. Nakmuenwai, P.; Yamazaki, F.; Liu, W. Automated Extraction of Inundated Areas from Multi-Temporal Dual-Polarizatio RADARSAT-2 Images of the 2011 Central Thailand Flood. *Remote Sens.* **2017**, *9*, 78. [CrossRef]

34. Bayik, C.; Abdikan, S.; Ozbulak, G.; Alasang, T.; Aydemir, S.; Sanli, F.B. Exploring multi-temporal Sentinel-1 SAR data for flood extend mapping. *Int. Arch. Photogramm. Remote Sens. Spatial Inf. Sci.* **2018**, *42*, 109–113. [CrossRef]
35. Martinis, S.; Kuenzer, C.; Wendleder, A.; Hult, J.; Twele, A.; Roth, A.; Dech, S. Comparing four operational SAR-based water and flood detection approaches. *Int. J. Remote Sens.* **2015**, *36*, 3519–3543. [CrossRef]
36. Pajuelas, G.B. A Study of Rainfall Variations in the Philippines: 1950–1996. *Sci. Diliman* **2000**, *12*, 1–28.
37. Cinco, A.T.; Hilario, D.F.; de Guzman, G.R.; Ares, D.E. Climate trends and projections in the Philippines. In Proceedings of the 12th National Convention on Statistics (NCS), Mandaluyong City, Philippines, 1–2 October 2013.
38. Torres, R.; Snoeij, P.; Geudtner, D.; Bibby, D.; Davidson, M.; Attema, E.; Potin, P.; Rommen, B.; Floury, N.; Brown, M.; et al. GMES Sentinel-1 mission. *Remote Sens. Environ.* **2012**, *120*, 9–24. [CrossRef]
39. Henry, J.B.; Chastanet, P.; Fellah, K.; Densos, Y.L. Envisat multipolarized ASAR for flood mapping. *Int. J. Remote Sens.* **2006**, *27*, 1921–1929. [CrossRef]
40. Schumann, G.; Bates, P.D.; Horritt, M.S.; Matgen, P.; Pappenberger, F. Progress in integration of remote sensing-derived flood extent and stage data and hydraulic models. *Rev. Geophys.* **2009**, *47*, RG4001. [CrossRef]
41. Brisco, B. Mapping and Monitoring Surface Water and Wetlands with Synthetic Aperture Radar. In *Remote Sensing of Wetlands: Applications and Advances*; Tiner, R., Lang, M., Klemas, V., Eds.; CRC Press: Boca Raton, FL, USA, 2015; pp. 119–136.
42. Chini, M.; Hostache, R.; Giustarini, L.; Matgen, P.A. Hierarchical Split-Based Approach for Parametric Thresholding of SAR Images: Flood Inundation as Test Case. *IEEE Trans. Geosci. Remote Sens.* **2017**, *55*, 6975–6988. [CrossRef]
43. Ulaby, F.T.; Dobson, M.C. *Handbook of Radar Scattering Statistics for Terrain*; Arthech House: Norwood, MA, USA, 1989; ISBN 0890063362.
44. Small, D. Flattening gamma: Ratiometric terrain correction for SAR imagery. *IEEE Trans. Geosci. Remote Sens.* **2011**, *49*, 3081–3093. [CrossRef]
45. Lee, J.S.; Wen, J.H.; Ainsworth, T.L.; Chen, K.-S.; Chen, A.J. Improved sigma filter for speckle filtering of SAR imagery. *IEEE Trans. Geosci. Remote Sens.* **2009**, *47*, 202–213.
46. Lee, J.S.; Jurkevich, I. Speckle Filtering of Synthetic Aperture Radar Images: A review. *Remote Sens. Rev.* **1994**, *8*, 313–340. [CrossRef]
47. Whoodhouse, I.H. *Introduction to Microwave Remote Sensing*; CRC Press: Boca Raton, FL, USA, 2017.
48. Shumann, G.; Di Baldassarre, G.; Bates, P. The utility of space-borne radar to render flood inundation maps based on multialgorithm ensembles. *IEEE Trans. Geosci. Remote Sens.* **2009**, *47*, 2801–2807. [CrossRef]
49. Manjusree, P.; Kumar, L.P.; Bhatt, C.M.; Rao, G.S.; Bhanumurthy, V. Optimization of threshold ranges for rapid flood inundation mapping by evaluating backscatter profiles of high incidence angle SAR Images. *Int. J. Disaster Risk Sci.* **2012**, *3*, 113–122. [CrossRef]
50. Long, S.; Fatoyinbo, T.E.; Policelli, F. Flood extent mapping for Namibia using change detection and thresholding with SAR. *Int. J. Environ. Res. Lett.* **2012**, *9*, 035002. [CrossRef]
51. Mason, D.C.; Speck, R.; Devereux, B.; Schumann, G.J.-P.; Neal, J.C.; Bates, P.D. Flood detection in urban areas using TerraSAR-X. *IEEE Trans. Geosci. Remote Sens.* **2010**, *48*, 882–894. [CrossRef]
52. Glasbey, C. An analysis of histogram-based thresholding algorithms. *CVGIP Graph. Models Image Process.* **1993**, *55*, 532–537. [CrossRef]
53. Fan, J.; Lei, B. A modified valley-emphasis method for automatic thresholding. *Pattern Recogn. Lett.* **2012**, *33*, 703–708. [CrossRef]
54. Al-Bayanti, M.; El-Zaart, A. Automatic thresholding techniques for SAR images. In Proceedings of the International Conference of Soft Computing, Dubai, UAE, 2–3 November 2013.
55. Martinis, S.; Twele, A.; Voigt, S. Towards operational near-real time flood detection using a split-based automatic thresholding procedure on high resolution TerraSAR-X data. *Nat. Hazards Earth Syst. Sci.* **2009**, *9*, 303–314. [CrossRef]
56. Martinis, S.; Kersten, J.; Twele, A. A fully automated TerraSAR-X based flood service. *ISPRS Int. J. Photogramm. Remote Sens.* **2015**, *104*, 203–212. [CrossRef]
57. Martinis, S.; Twele, A.; Strobl, C.; Kersten, J.; Stein, E. A Multi-scale flood monitoring system based on fully automatic MODIS and TerraSAR-X processing chains. *Remote Sens.* **2013**, *104*, 203–212. [CrossRef]

58. Hostanche, R.; Matgen, P.; Schumann, G.; Puech, C.; Hoffmann, L.; Pfister, L. Water level estimation and reduction of hydraulic model calibration uncertainties using satellite SAR images of floods. *IEEE Trans. Geosci. Remote Sens.* **2009**, *47*, 882–894.
59. Nguyen, B.D. Automatic detection of surface water bodies from Sentinel-1 SAR images using Valley-Emphasis methods. *Vietnam J. Earth Sci.* **2015**, *37*, 328–343.
60. Fuang, H.; Jargalsaikhan, D.; Tsai, H.-C.; Lin, C.-Y. An Improved Method for Image Thresholding based on the Valley-Emphasis Method. In Proceedings of the Asia-Pacific Signal and Information Processing Association Annual Summit and Conference, Kaohsiung, Taiwan, 29 October–1 November 2013. [CrossRef]
61. Otsu, N. A threshold selection method from gray-level histogram. *IEEE Trans. Syst. Man Cybern.* **1979**, *9*, 62–66. [CrossRef]
62. Zheng, Y.; Zhang, X.; Hou, B.; Liu, G. Using Combined Difference Image and k-Means Clustering for SAR Image Change Detection. *IEEE Geosci. Remote Sens. Lett.* **2014**, *11*, 691–695. [CrossRef]
63. Celik, T. Unsupervised Change Detection in Satellite Images Using Principal Component Analysis and k-Means Clustering. *IEEE Geosci. Remote Sens. Lett.* **2009**, *6*, 772–776. [CrossRef]
64. Ravichandran, K.S.; Ananthi, B. Color Skin Segmentation Using K-Means Cluster. *Int. J. Comput. Appl. Math.* **2009**, *4*, 153–157.
65. Santoro, M.; Wegmüller, U.; Lamarche, C.; Bontemps, S.; Defourny, P.; Arino, O. Strengths and weaknesses of multi-year Envisat ASAR backscatter measurements to map permanent open water bodies at global scale. *Remote Sens. Environ.* **2015**, *171*, 185–201. [CrossRef]
66. SWBD. Shuttle Radar Topography Mission Water Body Data Set. Digital Media. 2005. Available online: https://dds.Cr.Usgs.Gov/srtm/version2_1/ (accessed on 27 July 2019).
67. Gahlaut, S. Determination of Surface Water Area Using Multitemporal SAR Imagery. Master's Thesis, University of Stuttgart, Stuttgart, Germany, 2015. [CrossRef]

© 2019 by the authors. Licensee MDPI, Basel, Switzerland. This article is an open access article distributed under the terms and conditions of the Creative Commons Attribution (CC BY) license (http://creativecommons.org/licenses/by/4.0/).

Article

Improving Urban Flood Mapping by Merging Synthetic Aperture Radar-Derived Flood Footprints with Flood Hazard Maps

David C. Mason [1,*], John Bevington [2], Sarah L. Dance [3,4], Beatriz Revilla-Romero [2], Richard Smith [2], Sanita Vetra-Carvalho [5] and Hannah L. Cloke [1,3,6]

1. Department of Geography and Environmental Science, University of Reading, Reading RG6 6AB, UK; h.l.cloke@reading.ac.uk
2. JBA Consulting, Broughton Park, Skipton BD23 3FD, UK; John.Bevington@jbaconsulting.com (J.B.); Beatriz.Revilla-Romero@jbaconsulting.com (B.R.-R.); Richard.Smith@jbaconsulting.com (R.S.)
3. Department of Meteorology, University of Reading, Reading RG6 6ET, UK; s.l.dance@reading.ac.uk
4. Department of Mathematics and Statistics, University of Reading, Reading RG6 6AX, UK
5. Spire Global Ltd., Glasgow G3 8JU, UK; s.vetra-carvalho@reading.ac.uk
6. Department of Earth Sciences, Uppsala University, SE-751 05 Uppsala, Sweden
* Correspondence: d.c.mason@reading.ac.uk; Tel.: +44-118-378-8740

Citation: Mason, D.C.; Bevington, J.; Dance, S.L.; Revilla-Romero, B.; Smith, R.; Vetra-Carvalho, S.; Cloke, H.L. Improving Urban Flood Mapping by Merging Synthetic Aperture Radar-Derived Flood Footprints with Flood Hazard Maps. *Water* **2021**, *13*, 1577. https://doi.org/10.3390/w13111577

Academic Editors: Alberto Refice, Domenico Capolongo, Marco Chini and Annarita D'Addabbo

Received: 1 May 2021
Accepted: 28 May 2021
Published: 2 June 2021

Publisher's Note: MDPI stays neutral with regard to jurisdictional claims in published maps and institutional affiliations.

Copyright: © 2021 by the authors. Licensee MDPI, Basel, Switzerland. This article is an open access article distributed under the terms and conditions of the Creative Commons Attribution (CC BY) license (https://creativecommons.org/licenses/by/4.0/).

Abstract: Remotely sensed flood extents obtained in near real-time can be used for emergency flood incident management and as observations for assimilation into flood forecasting models. High-resolution synthetic aperture radar (SAR) sensors have the potential to detect flood extents in urban areas through clouds during both day- and night-time. This paper considers a method for detecting flooding in urban areas by merging near real-time SAR flood extents with model-derived flood hazard maps. This allows a two-way symbiosis, whereby currently available SAR urban flood extent improves future model flood predictions, while flood hazard maps obtained after the SAR overpasses improve the SAR estimate of urban flood extents. The method estimates urban flooding using SAR backscatter only in rural areas adjacent to urban ones. It was compared to an existing method using SAR returns in both rural and urban areas. The method using SAR solely in rural areas gave an average flood detection accuracy of 94% and a false positive rate of 9% in the urban areas and was more accurate than the existing method.

Keywords: image processing; hydrology; synthetic aperture radar

1. Introduction

Flooding causes significant death, injury, displacement, homelessness and economic loss all over the world every year. The risks to people and the economic impacts of flooding are greatest for urban flooding [1–4]. For example, regarding riverine floods in the UK, over 2 million properties (the majority of them in urban areas) are located in floodplains. An estimated 200,000 of these properties are at risk because they do not have protection against a 1-in-75-year flood event [5]. The number of floods and the number of properties affected by them are likely to increase in the future, due to the growing population exposure in floodplains and the impact of climate change [6]. In economically strong and populated areas, global economic losses due to floods are projected to reach $597 billion over the period 2016–2035 [7].

High-resolution SAR sensors are now commonly used for flood detection because of their ability (unlike visible-band sensors) to penetrate the clouds that are often present during flooding and to image at night as well as during the day. A number of very-high-resolution (VHR) SARs with spatial resolutions as high as 3 m or better are capable of detecting urban flooding, including TerraSAR-X, ALOS-2/PALSAR-2, the 3-satellite RADARSAT-2 constellation, the four satellites of the COSMO-SkyMed constellation, and the three satellites of the IceEye constellation [8,9]. In the absence of significant surface water turbulence caused by wind, rain or currents, floodwater generally appears dark in a SAR image due to specular reflection from the water surface away from the antenna. In addition, an attractive option for flood studies is the high-resolution (HR) Sentinel-1 constellation, which provides open-access satellite data at 10 m spatial resolution in near real-time, acquired according to a preplanned schedule. In common with RADARSAT-2, Sentinel-1 provides the user with processed multilook georegistered SAR images about one hour after image reception at the ground station.

An important use of the flood extent from a near real-time SAR image is as a tool for operational flood incident management [10]. Rapid response to flooding is essential to minimise loss of life and reduce suffering. Knowledge of the flooding situation is crucial for personnel deployment, resource allocation and rescue operations. The English Environment Agency (EA) now uses SAR images to detect the extent and depth of flooding as floods evolve [11]. The data may also be used for assimilation into urban flood inundation models, improving the model state and providing estimates of the model parameters and external forcing [12–19].

The problem of automated flood detection in rural areas has received a substantial amount of attention in the literature [20–36]. The Copernicus Emergency Management Service (EMS) and the EA are among several organisations that have developed semi-automatic systems to extract flood extents from a SAR image. Fully automated systems have also recently been developed using deep learning methods [35,36]. All these systems tend to work well in rural areas but have difficulty detecting urban flooding, primarily because SAR is a side-looking instrument. As a result, substantial areas of urban ground may not be visible to the SAR due to radar layover and shadow caused by buildings [37]. As shadows will appear dark, they may be misclassified as water if the ground in shadow is dry, whereas layover will generally appear bright and may possibly be misclassified as unflooded when, in reality, the ground is flooded. In addition, double scattering between ground surfaces and adjacent buildings often causes strong returns that confuse the image [37]. A further difficulty is that unflooded roads and tarmac areas also exhibit low backscatter, though often not as low as undisturbed water [25]. The dielectric constant of tarmac is considerably lower than that of water, and undisturbed water is smoother than tarmac, implying an increase of surface reflectivity and a consequent reduction in backscatter [38].

As a result of these difficulties, less attention has been given to research into urban flood detection using SAR. Despite this, several studies have now been performed that have employed a number of different techniques. These include analysing the backscatter returns in a post-flood SAR image [39–43], considering changes in backscatter intensities between pre- and post-flood SAR images [29], exploiting interferometric coherence, as well as backscatter intensities, using pre- and post-flood SAR images [8,38,44,45], and analysing SAR image time series [46].

An approach to urban flood detection that requires only pre- and post-flood Sentinel-1 imagery, allowing any flooding to be rapidly detected, is described in [44]. Strong double scattering in the pre-flood image from buildings roughly aligned with the satellite direction of travel is first used to detect urban areas [34,41,47]. Coherence changes between pre- and post-flood images are then used to refine the urban flooding determined using SAR intensity. Coherence should be high in urban areas that are not flooded, but low if there is flooding. The method achieved good results using imagery of flooding in Houston, Texas due to Hurricane Harvey in 2017. Li et al. (2019a) also employed pre- and post-flood Sentinel-1 data from the same flood to demonstrate that coherence provides valuable additional information to intensity in urban flood mapping, using an unsupervised Bayesian network fusion of intensity and coherence data [8]. In a similar manner, Li et al. (2019b) performed urban flood mapping with an active self-learning neural network based on TerraSAR-X intensity and coherence, again using the Houston flooding as a case study [45]. Good detection accuracy was obtained in areas containing fairly low-density detached suburban housing. These papers demonstrate significant progress towards urban flood detection using SAR data alone. Nevertheless, the flood detection accuracies obtainable in dense urban areas using HR SAR data were not fully explored. Many existing towns have much higher housing densities than the Houston suburbs (new estates in England are ~×8 denser). Mason et al. (2021) [48] describe a change detection method of detecting flooding in dense urban areas using Sentinel-1 and the WorldDEM Digital Surface Model (DSM) [49]. Flood levels in urban areas are estimated at double scatterers using increased SAR backscatter in the post-flood image due to double scattering between water and adjacent buildings compared to the unflooded case, wherein the double scattering is between ground and buildings. Areas of urban flooding are detected by comparing an interpolated flood level surface to the DSM.

Considering flood detection in urban areas of both high and low housing density using VHR SAR, Mason et al. (2018) used LiDAR data of the urban areas in conjunction with a SAR simulator to predict areas of layover and shadow in the image caused by buildings and taller vegetation [42]. Flooding was detected in urban areas not in shadow or layover by analysing the backscattered intensities from a single-polarisation VHR SAR image acquired during the flooding. Flooding detected in these areas was propagated into adjacent areas of shadow and layover, provided they were of similar elevation to the flooded areas, irrespective of their backscatter. Considering the percentage of the urban flood extent visible in the validation data that was detected by the SAR, the flood detection accuracy averaged over the three test examples studied was 79%, with a false alarm rate of 10%. The results indicated that flooding could be detected in urban areas with reasonable accuracy but also that this accuracy was limited by the VHR SAR's poor visibility of the urban ground due to shadow and layover and the backscatter similarity between urban floodwater and unflooded urban surfaces.

This paper considers the merits of an alternative method of improving the accuracy of rapid post-event flood mapping in urban areas by merging precomputed flood return period (FRP) maps with VHR SAR-derived flood inundation maps [43] and compares the method with that of [42]. Tanguy et al. (2017) mapped river flooding in urban areas using RADARSAT-2 backscatter intensities together with FRP data produced by a hydrodynamic model [43]. The flood level was estimated in rural areas using a post-flood SAR image, and this rural flood level was used with the FRP data to calculate where the flooding should be in the adjacent urban areas. A high accuracy of urban flood detection (87%) was achieved on the test cases studied, with a false alarm rate of 14%. Because the SAR data are used only in rural areas, the method has the advantages that there is no need to calculate shadow/layover maps in the urban areas, and that false alarms from unflooded urban surfaces with responses similar to water are eliminated. The advantage of the FRP maps is that, if there are urban areas that are either higher than the flooding or lower than it but defended (e.g., by embankments), this information should be contained in the maps, which would associate high return periods with such areas. A disadvantage is that it is necessary that the FRP data be accurate. The model predicts only flooding that is fluvial in origin, and it must be assumed that the rainfall pattern across the catchment that caused a particular flood is the same as that used to calculate the FRP maps. However, if a flood event was due to pluvial as well as fluvial flooding, the rural SAR water level observations should be able to correct errors in model water elevations in their immediate neighborhoods at least.

The paper extends that of [43] by estimating urban flooding in image sequences. The inputs are dynamic flood inundation extent and depth maps (updated every 3 h) produced by the Flood Foresight system [50], together with a contemporaneous SAR image sequence. Real-time flood modelling may often be carried out during a flood event, and the combination of the SAR and model flood extents may allow the model to be kept on track to make future predictions more accurately. The linking of the SAR and model data in this way allows a two-way symbiosis, whereby currently available SAR urban flood extent improves future model flood predictions, while flood hazard maps obtained after the SAR overpass improve the SAR estimate of the urban flood extent.

The present study and those of [42,43] require both VHR SAR and accurate high-resolution DSM data to be available. The method employs a LiDAR DSM, which limits its use to urban regions that have been mapped using airborne LiDAR. However, most major urban areas in flood plains in the UK and other developed countries have now been mapped. The hydrodynamic model used to estimate the FRP data also requires an accurate DSM, though this need not be as high-resolution as LiDAR.

The object of the paper is to investigate whether, given contemporaneous SAR, model FRP and flood inundation data, urban flooding can best be predicted using

(a) SAR in rural areas,
(b) SAR in rural areas and precomputed FRP maps in urban areas,
(c) SAR in rural areas and FRP maps and dynamic model flood inundation in urban areas,
(d) SAR in both rural and urban areas (with associated urban shadow/layover maps).

2. Materials and Methods

2.1. Flood Foresight System

The SAR flood footprints are compared to dynamic, event-specific flood inundation maps generated by JBA's Flood Foresight system [50]. Flood Foresight is an operational system developed to rapidly provide broad-scale estimates of flood hazard and impacts before, during, and after major riverine flood events (Figure 1).

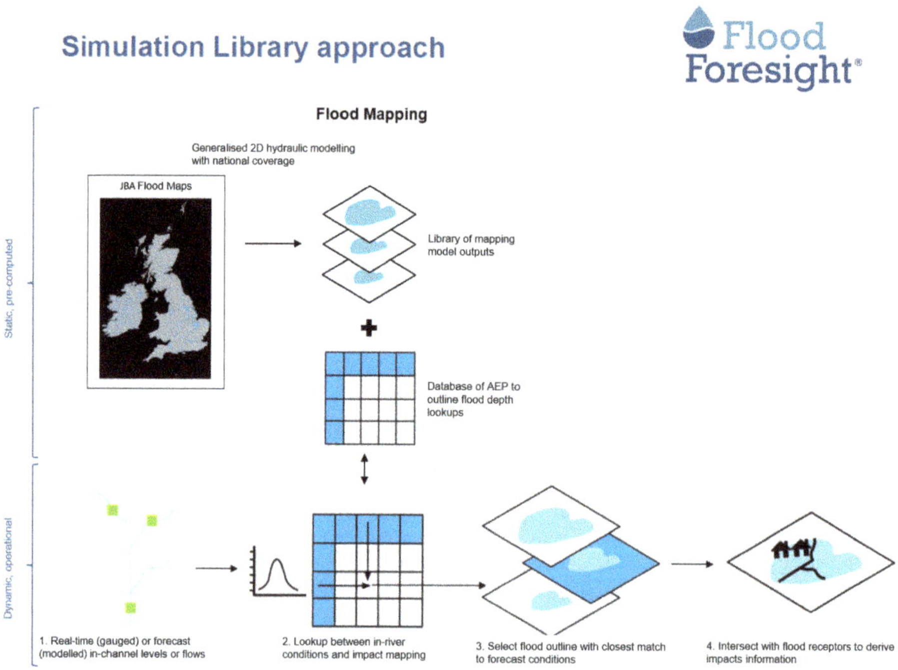

Figure 1. Simulation Library approach.

2.1.1. Flood Return-Period Maps

In the U.K., the JFlow flood inundation model [51] is used to produce a basic set of 5 m resolution flood hazard maps. These cover 1 in 20-, 75-, 100-, 200-, and 1000-year return periods (annual exceedance probabilities (AEPs) = 5%, 1.3%, 1%, 0.5%, and 0.1%, respectively).

2.1.2. Flood Foresight Model

The Flood Foresight system provides spatial data representing flood inundation spatial extents and depths with which users can measure their current or predicted impact from an event, an omission and major limitation of many current flood forecasting models [52]. Flood Foresight includes modules that provide both forecast (flood forecasting module) and real-time (flood monitoring module) flood inundation extents and depth data, driven by a range of forecast or telemetered streamflow data. The flood forecasting module provides daily forecasts of flooding up to 10 days in advance by linking forecast river flow data from European-scale and global hydrological models to an enhanced 'simulation library' of precomputed FRP maps (Section 2.1.1). The flood monitoring module (the component that generated the data used in this study) provides similar output of flood inundation data but in near real-time — every 3 h as an event unfolds — by combining the enhanced FRP map library with observed river gauge telemetry for England, Scotland and Wales from the Environment Agency (EA), Scottish Environment Protection Agency (SEPA) and Natural Resources Wales (NRW), respectively.

A simulation library containing precomputed FRP maps is employed for flood mapping activity (Figure 1). The simulation library approach used within Flood Foresight provides a solution to the problem of generating national-scale flood footprints from hydraulic models in near real-time or within a reasonable timescale in order for the data to be of use to flood forecasters and decision-makers. The simulation library method was one of two methods identified by the EA to provide national-scale flood inundation mapping capability [52]. Using these methods, Flood Foresight is able to generate a national-scale estimate of flood inundation across Great Britain (GB) in less than 10 min, thus giving the performance required for a national-scale strategic flood warning system.

The Flood Foresight system uses a set of downscaled (30 m) FRP maps, based on those data described in Section 2.1, as the basis for its simulation library. For each pair of contiguous maps, an interpolation technique is applied to derive 5 intermediate maps equally spaced in the return period. The interpolation approach used computes five intermediate depth grids using a depth-slicing algorithm to achieve approximate interpolation: the volume of water on the floodplain is first computed for each modelled RP, building a volume versus RP (VRP) curve for every 1 km^2 cell of floodplain. The algorithm then slices the vertical depth differences between consecutive modelled RP depth grids into depth intervals, generating intermediate depth grids.

To generate contiguous flood maps for each timestep, Flood Foresight calculates the current RP from a flow prediction (flood forecasting module) or gauge (flood monitoring module) and uses the VRP curve to estimate what volume would be expected. The interpolated depth grid with the closest volume to this is then selected for associated floodplain cells. Of course, the optimal method for developing intermediate FRP maps would be to run additional scenarios in the JFlow hydrological model. However, running this would be expensive at a national or global scale.

Thus, the simulation library used within Flood Foresight contains 30 return periods. A similar interpolation procedure was used in [43], where it was adopted because only 3–5 FRP maps are usually made publicly available, though it was admitted that using only a limited set of maps could lead to less reliable RP estimates. In Figure 1, the river gauge telemetry from the flood monitoring module of Flood Foresight is linked to the simulation library lookup tables, from which the flood return map with the closest match for the observed streamflow is selected.

2.2. Study Events and Data Sets

Four different SAR images of two different flood events were studied. The locations of the three study sites in southern Britain (Wraysbury, Staines, and Tewkesbury) are shown in Figure 2 [42]. The sites were chosen because they were subject to recent urban flood events for which VHR SAR, LiDAR, and independent validation data were available. Wraysbury is a village on the Thames, west of London; Staines is a town on the Thames nearer London with a higher housing density, and Tewkesbury is a market town on the Severn and Avon in the west of England.

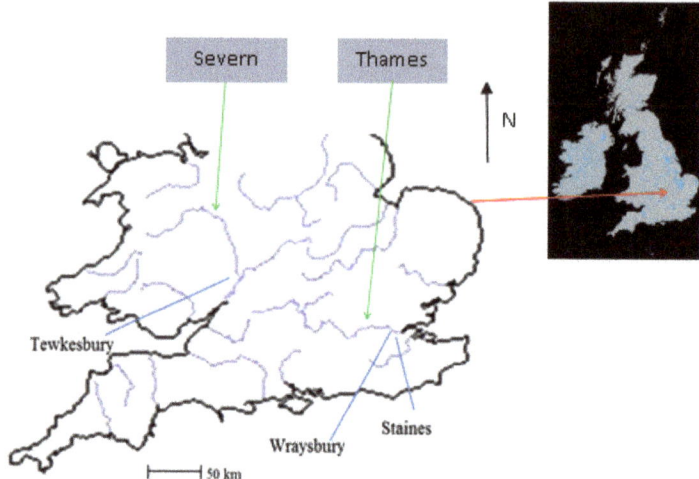

Figure 2. Locations of the 3 study sites in southern England, at Wraysbury (51.5° N, 0.6° W), Staines (51.4° N, 0.5° W) and Tewkesbury (52° N, 2.2° W) (main rivers in blue).

As stated in Mason et al. (2018) [42], "The first two examples are based on the Thames flood of February 2014 in West London, which caused substantial urban flooding [53]. In January and February 2014, heavy and persistent rainfall left large parts of southern England under water. The flooding resulted from a long series of Atlantic depressions caused by the jet stream being further south than usual. The peak of the flooding in West London occurred around 11 February 2014, with peak flow being 404 m^3/s. A substantial amount of urban flooding occurred in a number of towns, in particular Wraysbury and Staines. Three COSMO-SkyMed (CSK) (X-band) 2.5 m resolution Stripmap images of the flooding were acquired covering the flooded areas. Their processing level was GTC (Level 1D). A limited number of aerial photos acquired by the press were available to validate the SAR flood extents. These tended to cover small areas with substantial flooding. An example aerial photo showing flooding in Wraysbury is shown in Figure 3a, together with the SAR subimage for 12 February 2014 covering the area (Figure 3b). No high resolution visible band satellite (e.g., WorldView-2) data with low cloud cover were available for validation. The data acquired for the Thames flood were –

(a) a sequence of 3 CSK images showing flooding in the Wraysbury area on 12, 13 and 14 February 2014 just after the flood peak.
(b) a sequence of 2 CSK images on 13 and 14 February 2014 also showing flooding in Staines, where on 13 February the flow was still only 5% less than the peak. A contemporaneous aerial photo for validation was acquired showing flooding in Blackett Close, Staines.

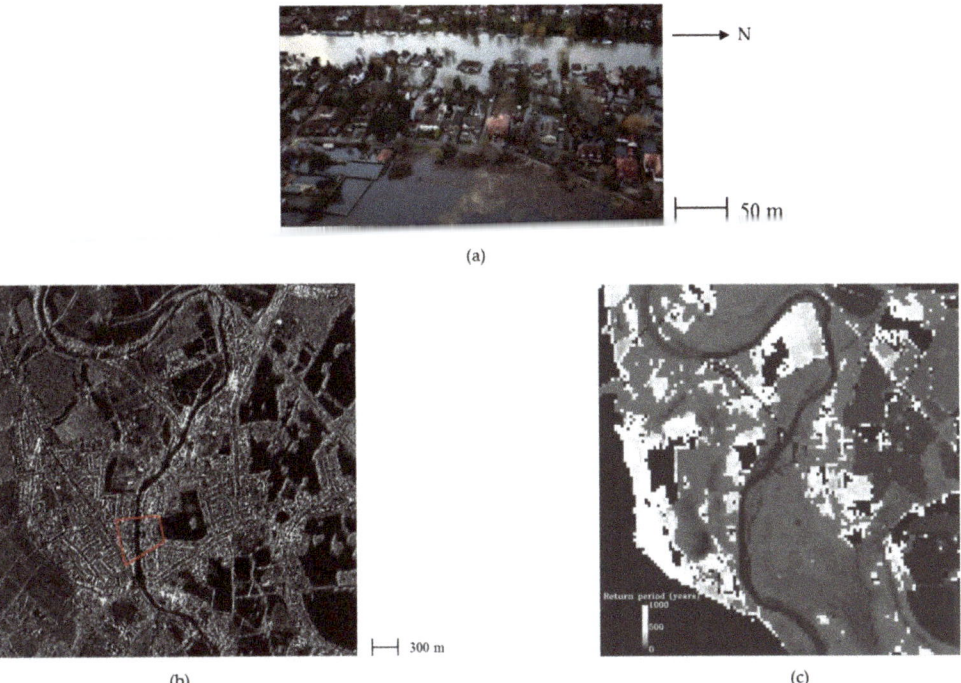

Figure 3. (**a**) Aerial photo of flooding in Wraysbury, West London (51.5° N, 0.6° W; about 300 × 300 m) (© Getty Images 2014) (after [38]), (**b**) CSK subimage (3 × 3 km) of Thames flood in Wraysbury, West London on 12 February 2014 (pixel intensities are digital number (DN) backscatter values; dark areas are water; red outline shows the area covered by aerial photo), and (**c**) flood return-period map (black areas masked out).

The third example was based upon the >1-in-100-year flood (AEP < 1.0%) that took place on the lower Severn around Tewkesbury in July 2007 [54]. This resulted in substantial flooding of urban and rural areas, about 1500 homes in Tewkesbury being flooded. Tewkesbury lies at the confluence of the Severn, flowing in from the northwest, and the Avon, flowing in from the northeast. The peak of the flood occurred on 22 July, and the river did not return to bankfull until 31 July. On 25 July, TerraSAR-X (TSX) (X-band) acquired a 3 m resolution StripMap image of the region in which urban flooding was visible. The image was multi-look ground range spatially enhanced [39]. Aerial photos of the flooding were acquired on 24 and 27 July, and these were used to validate the flood extent extracted from the TerraSAR-X image [39]." The Tewkesbury event is included as a test of how well the FRP method performs at river confluences. Traditional flood risk management methods have typically sidestepped the issue of tributary dependence by focusing on modelling the T-year flow using a single water course. However, at river confluences, it is difficult to define a single event that will be exceeded once every T years, due to the multiple possible combinations of flow magnitude and timing on the tributaries [55].

Table 1 gives the parameters of the SAR images considered in the study. All images were HH (horizontal transmit, horizontal receive) polarization, which, for flood detection, is preferable to vertical or cross polarization because it gives the highest contrast between open water and unflooded regions [56]. For each area, the EA LiDAR DSM and 'bare-earth' digital terrain model (DTM) of the area were obtained at 2 m resolution.

Table 1. Parameters of SAR images.

Date and Time	River	Location	SAR	Resolution (m)	Pass	Angle of Inclination (°)	Angle of Incidence (°)
12 February 2014 19:05	Thames	Wraysbury	COSMO-SkyMed	2.5	Descending	97.9	43.4
13 February 2014 18:11	Thames	Wraysbury Staines	COSMO-SkyMed	2.5	Descending	97.9	31.6
14 February 2014 18:05	Thames	Wraysbury Staines	COSMO-SkyMed	2.5	Descending	97.9	35.9
25 July 2007 06:34	Severn/Avon	Tewkesbury	TerraSAR-X	3.0	Descending	97.4	24

2.3. Method

Steps in the processing chain for urban flood delineation are shown in Figure 4. These include preprocessing operations carried out prior to SAR image acquisition and Flood Foresight model output, and near real-time operations carried out after the georegistered SAR image and model output have been obtained. As stated in Mason et al. (2018) [42], "for the SAR data, the approach involves first detecting the flood extent in rural areas, and then detecting it in adjacent urban areas using a secondary algorithm guided by the rural flood extent. A rural area is considered to be one not significantly affected by shadow and layover. Note that this means that the method will not work in a situation where a flood is totally contained within an urban area. But even in a city, rural areas (e.g., parks) can often be found not far away from urban ones."

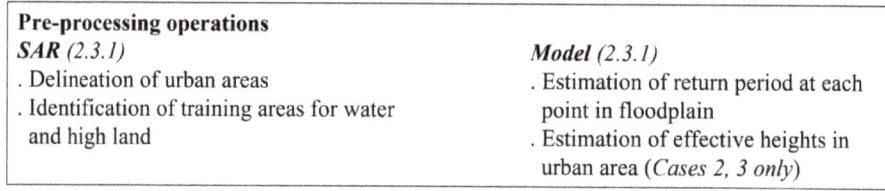

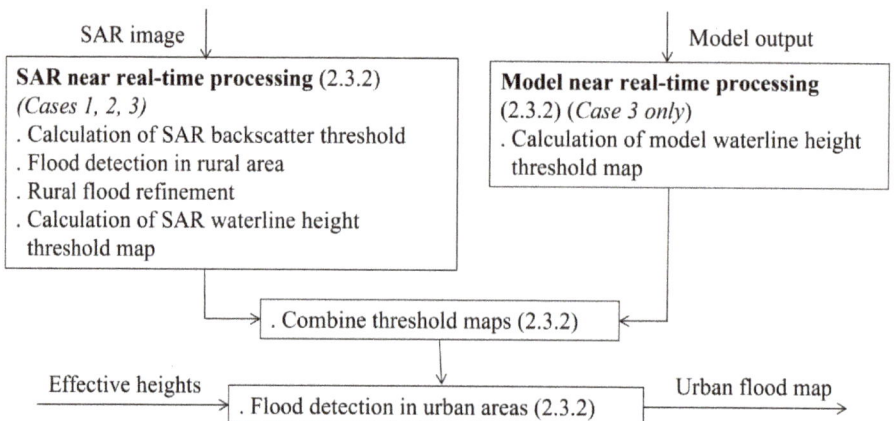

Figure 4. Steps in the processing chain for urban flood delineation (numbers in brackets refer to relevant section numbers; case numbers in italics; brackets refer to cases considered in the Results Section).

2.3.1. Preprocessing Operations
SAR

(a) Delineation of Urban Areas

Currently, the main urban areas are delineated manually, as this is a preprocessing operation that is not time-critical. Alternatively, the World Settlement Footprint 2015 10 m resolution dataset could be used to identify urban areas [57]. This is an open access dataset available on the European Space Agency Urban Thematic Exploitation Platform.

(b) Identification of Training Areas for Water and High Land

Training areas for the water and high-land classes are used to determine a SAR backscatter threshold to discriminate between flooded and unflooded areas. Unassigned heights in the LiDAR data, where the water has acted as a specular reflector reflecting too large a signal directly back to the LiDAR sensor, are used as the water training area. Unassigned heights may be present in unflooded river channels or permanent water bodies. The high land, which is not likely to be flooded, is taken as the highest 10% of pixels in the area. These pixels must not contain unassigned heights so that they are not water. Figure 9 of [42] shows an example of the training areas for the Wraysbury test site.

Flood Return-Period Maps

(a) Estimation of return period at Each Point in the Floodplain

A map $rp(x,y)$ giving the flood return period at each point (x,y) in the floodplain is generated using the flood inundation extents contained in the set of 30 binary FRP maps. The computation loops over the 30 maps in order of increasing return period. For a given FRP map, if a pixel (x,y) is non-zero and $rp(x,y)$ has yet not been set, $rp(x,y)$ is set to the return period of the current map. An example return-period map is shown in Figure 3c for the Wraysbury area.

(b) Estimation of Effective Heights in Urban Area

An effective height map $eff\_h(x,y)$ is determined in the urban area, given $rp(x,y)$ and the set of 30 binary FRP maps. In cases wherein the urban area is undefended, the effective height map is simply the DSM. In cases wherein some portion of the urban area is defended, this will be encapsulated in the FRP maps as high return periods at the defended pixels, which will result in these pixels having effective heights higher than the DSM heights at these pixels.

To take into account the fall-off of water level down the reach, and that different parts of a domain may be flooded to different depths, the domain is subdivided into nonoverlapping rectangular $m \times n$-pixel subdomains of ~1 km side. For each FRP map in each subdomain, edges are detected in the binary image using the Sobel edge detector. The edges are overlain on the DSM to obtain water level observations (WLOs) at the edges. The mean WLO is calculated for those edges that are not close to high slopes in the DSM, as the WLO of an edge is likely to be determined more accurately on a low slope [42]. For each subdomain, this results in a look-up table associating a mean WLO with each return period. For each return period, the mean WLOs in the subdomains are interpolated over the whole domain using bilinear interpolation.

The interpolated maps of mean WLOs for each return period are then converted to effective heights in the urban areas. Given the return period of a pixel $rp(x,y)$, if its mean WLO is less than the DSM and the DSM is more than 1 m higher than the DTM, $eff\_h(x,y)$ is set to the DSM height at the pixel, as the pixel is likely to coincide with a building or taller vegetation. If the mean WLO is ≤ 1 m higher than the DSM height at the pixel, $eff\_h(x,y)$ is set to the DSM height. Otherwise, $eff\_h(x,y)$ is set to the mean WLO for the pixel's return period, ensuring that defended urban areas have higher return periods.

Note that, if only SAR data exists and there are no model results, the flow diagram of Figure 4 is still valid if the effective height map simply becomes the DSM.

2.3.2. Near Real-Time Operations
SAR

Near real-time SAR processing can begin as soon as the processed georegistered SAR image becomes available. Each SAR image is subjected to the following steps.

(a) Calculation of SAR Backscatter Threshold

The SAR backscatter threshold that best separates the backscatter values of the water and high land pixels in the training classes is calculated using the Bayes minimum misclassification rule. Equal prior probabilities are assumed for each class. The threshold selected is the backscatter value T_u, giving the minimum misclassification of water and high-land (nonwater) pixels [40].

(b) Flood Detection in Rural Areas

As stated in Mason et al. (2012b) [40], "flood detection in rural areas is object-based and adopts the approach of segmenting the SAR image into regions of homogeneity and then classifying them, rather than classifying each pixel independently using a per-pixel classifier. The use of segmentation techniques provides a number of advantages compared to using per-pixel classification." The approach employed for rural flood detection in [20,21] is adopted, which involves segmentation and classification using the eCognition Developer software (Trimblr Geospatial, Munich; Germany) [58]. Regions of homogeneous SAR backscatter are detected using the multi-resolution segmentation algorithm, and all resulting rural regions with mean backscatter less than the threshold are classed as 'flood'. Details are given in [42].

eCognition Developer is also used to refine the initial rural flood segmentation by using different rules. For example, the backscatter threshold may be raised to include in the flood category regions of rural flooding adjacent to a flooded region that have slightly higher mean backscatter than the threshold T_u (e.g., due to wind ruffling the water surface in more exposed parts of the floodplain). Again, details are given in [42], Figure 11 of which gives a refined rural flood classification for the Wraysbury area for 12 February 2014.

(c) Calculation of Local Waterline Height Threshold Map

A local waterline height threshold map is calculated using the rural flood map. The method is based on the assumption that water in the urban areas should not be at a substantially higher level than in the surrounding rural areas. Waterline heights are calculated at positions of low slope in the DSM. The method also requires knowledge of the positions of permanent water bodies. In this case, high-resolution LiDAR data must be available, and, as any region imaged by LiDAR will generally also have land cover data available, permanent water bodies are extracted from a land cover map. However, they could also be extracted from a preflood SAR image.

As in the determination of effective heights (Section 2.3.1), the domain is subdivided into nonoverlapping rectangular $m \times n$-pixel subdomains, to take into account the fall-off of water level down the reach and the fact that different parts of a domain may be flooded to different depths. In each subdomain, as stated in Mason et al. (2018) [42], "waterlines are detected by applying the Sobel edge detector to the binary flood extent map. Because the flood map has errors at this stage, edges will be present at the true waterlines, but also in the interior of the water objects due to regions of emergent vegetation and shadow/layover (giving water heights that are too low), as well as above the waterline due to higher water false alarms. To increase the signal-to-noise ratio of true edges, a dilation and erosion operation is performed on the water objects to eliminate some of the artefacts [42]. Water objects are first dilated by 12 m, then eroded by the same amount. It is required that an edge pixel is present at the same location within a 2-m-wide buffer before and after dilation and erosion. The buffer is required because an edge that has been dilated and eroded may be smoother than the original edge, and may be slightly displaced from it as a result. This tends to select for true waterline segments on straighter sections of exterior boundaries of water objects. To suppress false alarms further, waterline heights in regions that are sufficiently far (11 m) from high (>0.5) DSM gradients are selected, provided that they are also within ±1.5 m of the mean water height. This avoids false alarms near high DSM slopes, which may give rise to shadow/layover areas [42]. At this stage also, waterlines from permanent water bodies are excluded using the land use map.

In order to find the mean waterline height in the rural area in each sub-domain, a histogram is constructed of the waterline heights, and the positions of the histogram maxima are found. Generally, the mean waterline height in the sub-domain is set to correspond to the height of the largest maximum. However, if any substantial maxima greater than half that of the largest maximum are present at a higher waterline height, the highest of these is chosen instead. This latter rule copes with the situation where a substantial number of erroneous low waterline heights in the interior of water objects have not been eliminated, leading to a largest maximum at an incorrect low height. An example histogram is shown in Figure 6 of [40]." A standard deviation for the mean waterline height is estimated using the histogram frequencies lying above it. A further check is carried out in the case where the domain is divided into two subdomains, one in the upper and one in the lower part of the reach. Occasionally it may happen that, in one of the subdomains, the maximum value h_1 chosen from the histogram is less than the overall mean calculated not from the histogram but directly from all the waterline heights in the subdomain w_1. In this case, h_1 is corrected using the value in the other subdomain (h_0) and the difference between the mean waterline values in the upper and lower domains ($w_0 - w_1$), i.e., $h_1 = h_0 - (w_0 - w_1)$. The waterline heights in the subdomains are then interpolated over the whole domain using bilinear interpolation.

A small positive guard height may be added to cope with bias introduced by the waterline heights not including the height of any flooded vegetation at the flood edge. The rural SAR segmentation algorithm does not take into account the fact that there may be emergent vegetation at the flood edge that will not be classed as flooded due to the high backscatter it produces [40]. Because of the difficulty of estimating this (usually short) vegetation height, the guard height is treated as a free parameter that must be optimized by calibration.

Flood Foresight Model Output

Near real-time operations also take place on the Flood Foresight model flood extent valid at the current time, in order to produce an analogous model waterline height threshold map. A similar procedure to that used in the determination of effective heights (Section 2.3.1) is followed. The domain is subdivided into nonoverlapping rectangular $m \times n$-pixel subdomains, to take into account the fall-off of water level down the reach and the fact that different parts of a domain may be flooded to different depths. In each subdomain, edges are detected in the binary model flood extent image using the Sobel edge detector. The edges are overlain on the DSM to obtain WLOs at the edges. The mean WLO and its standard deviation for the subdomain is calculated for those edges that are not close to high slopes in the DSM, as the WLO of an edge is likely to be determined more accurately on a low slope. It seems reasonable with model data to measure the mean WLO in each subdomain rather than searching for the highest maximum in the histogram of waterline heights as with the SAR WLOs, which may contain water levels that are too low (see previous Section). The mean WLOs in the subdomains are then interpolated over the whole domain using bilinear interpolation.

Near Real-Time Combined Processing

(a) Combination of Waterline Height Threshold Maps

At this stage the interpolated waterline height maps from the SAR and the model are combined to form a single waterline height map. A possible approach to this would be to assimilate a sequence of SAR WLO maps into the model WLO map as it evolves over time in order to improve the latter, perhaps using a sequential ensemble Kalman filter [13,14]. While this may be the optimum solution, in this case there is a difficulty in updating the model WLO map with height innovations because the simulation library approach used in the modelling means that, at this stage, there is no hydrodynamic model being run into which to assimilate the innovations.

The simpler approach adopted here is to use the best linear unbiased estimate [59]. Consider the simple weighting scheme

$$O(x,y,t) = (w_1 O_{SAR}(x,y,t) + w_2 O_{MOD}(x,y,t))/(w_1 + w_2) \qquad (1)$$

where $O_{SAR}(x,y,t)$ and $O_{MOD}(x,y,t)$ are SAR and model water-level observations at position (x,y) and time t, w_1 and w_2 are weights, and $O(x,y,t)$ is the combined estimate. Suppose $O_{SAR}(x,y,t)$ and $O_{MOD}(x,y,t)$ are independent unbiased estimators with variances σ_1^2 and σ_2^2, respectively. Then, if $w_1 = 1/\sigma_1^2$ and $w_2 = 1/\sigma_2^2$, $O(x,y,t)$ is the linear combined unbiased estimate with the minimum variance. The variance of $O(x,y,t)$ is

$$\sigma_1^2 \sigma_2^2 / (\sigma_1^2 + \sigma_2^2) \qquad (2)$$

The combined variance is lower than either σ_1^2 or σ_2^2, resulting in an improvement in accuracy, though this result holds only if the variances are known.

Rural SAR WLOs may well be more accurate than model WLOs at SAR acquisition time. But as time passes, the SAR flood extent will become more out-of-date until the next SAR image is acquired, whereas the model will be updated every 3 h, so the relative accuracies of SAR and model data will change with time. A simple way of taking this into account is by using an exponential "forgetting factor" and setting

$$w_1 = exp(-(t - t_0)/\tau)/\sigma_1^2 \qquad (3)$$

where t_0 is the time of the last SAR acquisition, and τ is the nominal lifetime of this SAR image's usefulness. As an example, assuming $\sigma_1 = 0.3$ m, $\sigma_2 = 0.4$ m and $\tau = 2$ days, then at SAR image acquisition time ($t = t_0$) around flood peak, $w_1/(w_1 + w_2) = 0.64$ and $w_2/(w_1 + w_2) = 0.36$, whereas at $t = 4$ days, $w_1/(w_1 + w_2) = 0.2$ and $w_2/(w_1 + w_2) = 0.8$.

(b) Flood Detection in Urban Areas

The final stage is to extract urban water regions from the effective height image $eff\_h(x,y)$ produced from the model return-period maps, using the combined SAR and model WLO threshold map $O(x,y,t)$. For pixels in urban areas, if

$$eff\_h(x,y) < O(x,y,t) \qquad (4)$$

then the pixel is flooded, else not flooded.

2.3.3. Performance Measures

The performance measures used to assess the flood detection accuracy were the flood detection rate (i.e., recall (or hit rate) = $t_p/(t_p + f_n)$), the precision (= $t_p/(t_p + f_p)$), and the critical success index (CSI = $t_p/(t_p + f_p + f_n)$), where t_p = true positives, f_n = false negatives, and f_p = false positives.

3. Results

The approach taken in the validation of the urban flood extents was to first test the method using just the rural SAR data and then to include the model data also to examine the manner and extent to which these could improve the results.

Flood extents for use as validation data were extracted from the aerial photos obtained contemporaneously with the SAR data in the three study areas as described in [40]. A difficulty with the Thames 2014 flood data was that the aerial photos were acquired after the SAR imagery, on 16 February 2014. However, the flooding was long-lasting, and data from the Staines flood gauge indicated that the river level had fallen only 20 cm in the intervening period since 12 February 2014 and 13 February 2014 and 10 cm since 14 February 2014. The mean waterline height was raised to compensate for this.

The guard height to be added to the SAR waterline height was calibrated at 0.4 m by minimizing the percentage of pixels misclassified (i.e., the sum of the false negative and false positive percentages) averaged over the three test sites studied.

3.1. Case 1: Results Using Rural SAR Data Only

Case 1 only used rural SAR WLOs in the method. This meant that the waterline height threshold map used in Figure 4 was derived solely from the SAR waterline map, and the effective height map was simply the DSM. The precomputed FRP maps and dynamic Flood Foresight model flood extents were not used. Table 2 gives the flood detection and false-alarm rates for the six SAR scenes.

Table 2. Urban flood detection accuracy using rural SAR WLOs only.

Image	Flood Detection Rate (Recall) (%)	Precision (%)	Critical Success Index (CSI) (%)
Wraysbury 12 February 2014	91	98	89
Wraysbury 13 February 2014	95	96	91
Wraysbury 14 February 2014	90	99	89
Blackett 13 February 2014	100	85	85
Blackett 14 February 2014	99	99	99
Tewkesbury 25 July 2007	88	77	70

For Wraysbury, the 3 × 3 km domain was divided into upper and lower windows (each 3 × 1.5 km) along the reach, to take account of the fall-off down the reach. Figure 5 shows the correspondence between the SAR and aerial photo flood extents in the Wraysbury validation area for the three SAR scenes of 12–14 February 2014, together with an extract from each SAR image for comparison. An average flood detection rate of 92% was achieved for the Wraysbury images, together with an average precision of 98%.

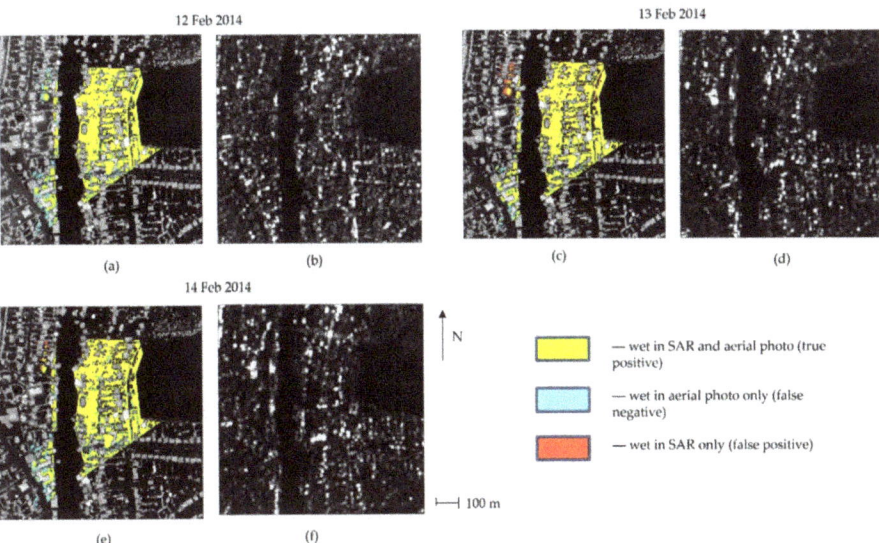

Figure 5. (a) Correspondence between the SAR image of 12 February 2014 and aerial photograph flood extents in urban area of Wraysbury, superimposed on the LiDAR image (lighter grey = higher), and (b) extract from SAR image of 12 February 2014; (c,d) as (a,b) for SAR image of 13 February 2014; and (e,f) as (a,b) for SAR image of 14 February 2014.

For Blackett Close, the 3 × 3 km domain was divided into upper and lower windows as for Wraysbury. Figure 6 shows the aerial photo used for validation, the SAR subimage for 13 February 2014 covering this, and a return-period map for the Staines area. Figure 7 shows the correspondence between the SAR and aerial photo flood extents in the Blackett Close validation area for the SAR scenes of 13–14 February 2014 together with an extract from each SAR image for comparison. Very high flood-detection accuracies were obtained in this validation area, with an average flood detection rate of 99.5%, though with a slightly higher average false positive rate leading to a lower average precision of 92%. However, it should be noted that the validation areas for Blackett Close and Wraysbury were limited and contained housing that was not particularly dense.

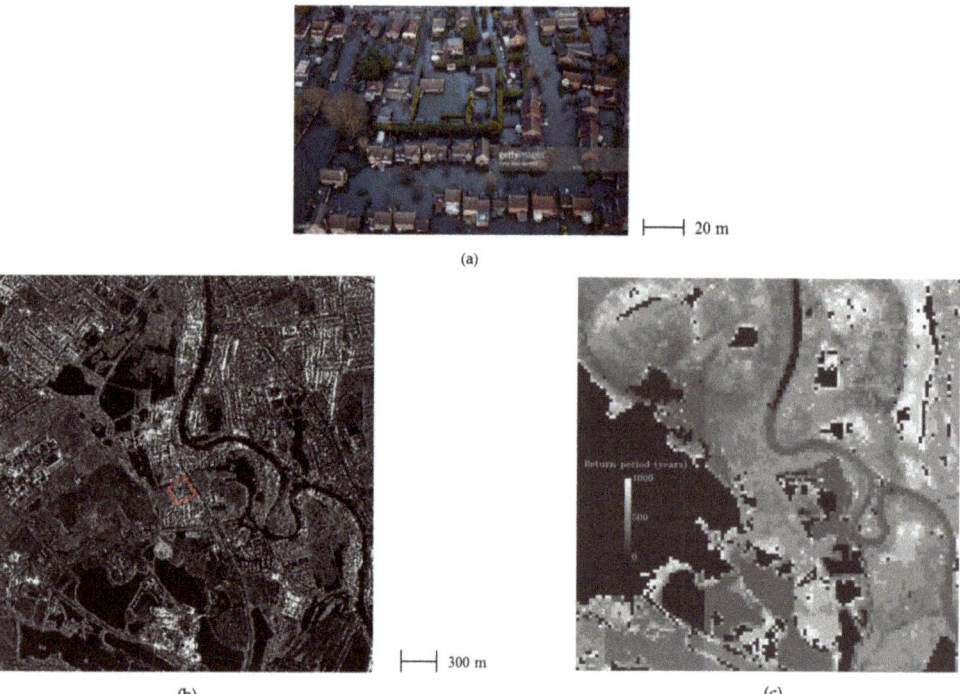

Figure 6. (**a**) Aerial photo of flooding in Blackett Close, Staines (51.4° N, 0.5° W; about 150 × 150 m) (© Getty Images 2014) (after [42]), (**b**) CSK subimage (3 × 3 km) of Thames flood in Staines, West London on 13 February 2014 (pixel intensities are DN backscatter values, and dark areas are water;red outline shows the arecovered by aerial photo), and (**c**) flood return-period map (black areas masked out).

For Tewkesbury, the 2.6 × 2 km domain was divided into four windows, each of 1.3 × 1 km, to reflect the fact that Tewkesbury lies on the confluence of the Severn and the Avon. Figure 8 shows the SAR subimage of 25 July 2007 covering the urban areas together with the associated return-period map. The extensive aerial photography used for validation on that date is shown in Figure 3 of [39], which also describes the steps used to process the validation data. Figure 9 shows the correspondence between the SAR and aerial photo flood extents in the Tewkesbury urban area. A flood detection rate of 88% was achieved, though the false positive rate was rather high, leading to a precision of 77%.

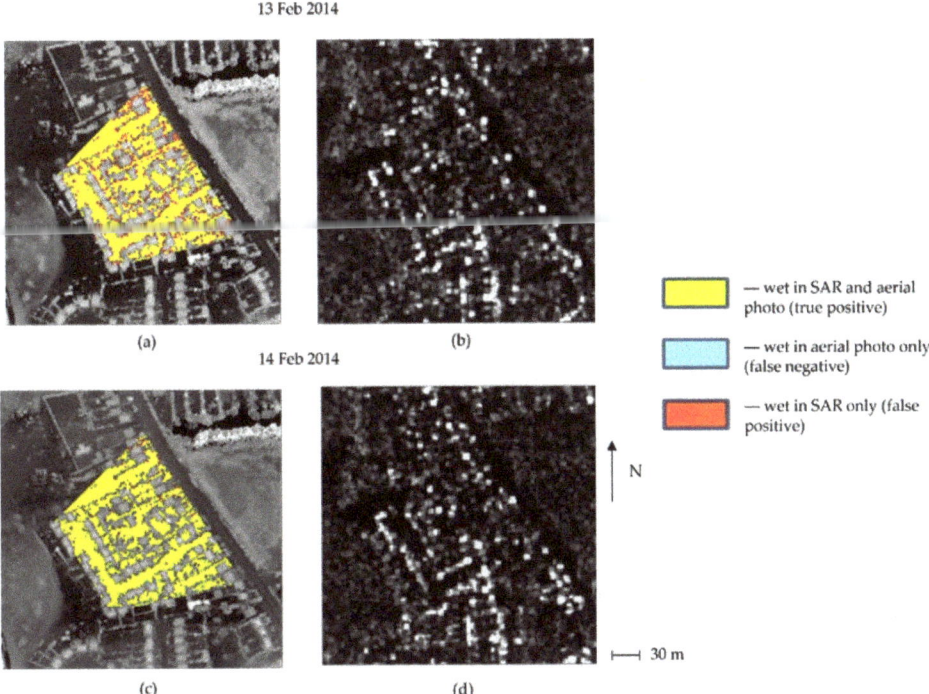

Figure 7. (**a**) Correspondence between the SAR image of 13 February 2014 and aerial photograph flood extents in the urban area of Blackett Close, superimposed on the LiDAR image (lighter grey – higher) and (**b**) extract from the SAR image of 13 February 2014; (**c**,**d**) as (**a**,**b**) for the SAR image of 14 February 2014.

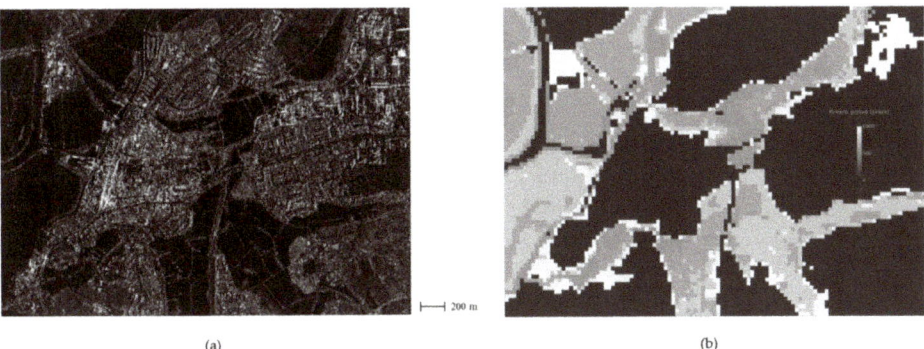

Figure 8. (**a**) TerraSAR-X image showing flooding in the Tewkesbury area on 25 July 2007 (52° N, 2.2° W, pixel intensities are DN backscatter values, and dark areas are water, 2.6 × 2 km, © DLR, after [39]), and (**b**) flood return-period map (black areas masked out).

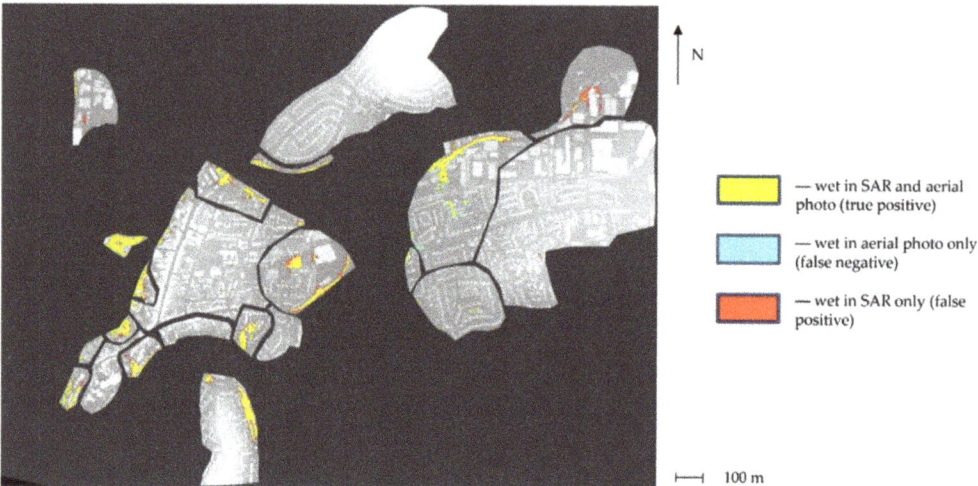

Figure 9. Correspondence between the SAR and aerial photograph flood extents in the urban area of Tewkesbury, superimposed on the LiDAR image (lighter grey – higher).

Averaged over all six SAR subimages, the flood detection rate was 94% and the precision was 92%.

As Case 1 used only SAR and not model data, it allowed a comparison of the present method with that described in [42]. The latter used the SAR data to detect flooding in the urban, as well as the rural, area and employed a SAR simulator in conjunction with LiDAR data of the urban area to predict areas of radar shadow and layover in the image caused by buildings and taller vegetation. Three of the SAR subimages used in the present study were also employed in the previous one, namely Wraysbury 12 February 2014, Blackett Close 13 February 2007 and Tewkesbury 25 July 2007. Considering the percentage of the urban flood extent visible in the aerial photo that was detected by the SAR, in the previous study, the average flood detection rate was 79%, with a precision of 89%. Mason et al. (2018) [42] concluded that "flooding could be detected in the urban area by this method to good, but perhaps not very good, accuracy, partly because of the SAR's poor visibility of the ground surface due to shadow and layover." For the same three subimages using the present Case 1 method, the average flood detection rate was 93%, and the precision was 86%, giving a lower overall error. As a result, the present method seems both more accurate than the previous method and simpler to implement, as no SAR simulation of shadow and layover in the urban area is required.

3.2. Case 2: Results Using Rural SAR Data and Pre-Computed FRP Maps

In this case, the waterline height threshold map of Figure 4 was again derived solely from the rural SAR waterline map (so that the dynamic Flood Foresight model flood extents were again not used), but the effective height map was derived from the model's FRP maps together with the DSM. Case 2 is the one most similar to that of [43]. The domains were divided as for Case 1. The results for all SAR scenes are shown in Table 3 and are very similar to those of Table 2, wherein only SAR data were employed to detect urban flooding.

Table 3. Urban flood detection accuracy using rural SAR WLOs and model FRP maps.

Image	Flood Detection Rate (Recall) (%)	Precision (%)	Critical Success Index (CSI) (%)
Wraysbury 12 February 2014	91	98	89
Wraysbury 13 February 2014	95	96	91
Wraysbury 14 February 2014	89	99	88
Blackett 13 February 2014	100	85	85
Blackett 14 February 2014	99	99	99
Tewkesbury 25 July 2007	88	77	70

An advantage of the flood return-period maps should be that they contain information on defended regions in the urban areas. On a practical note, the Flood Foresight system provides data on flood defences in a separate layer to the FRP maps, though the layers can be combined when forming the effective height map. However, there were no examples of defended regions in any of the test areas, and as a result, the effective height map for each area was very similar to its DSM, which explains the similarity of the results for Case 1 and Case 2. To illustrate the potential advantage of using FRP maps, it was necessary to simulate a defended region. Figure 10a shows the flood return map for the Staines area when the aerial photo validation area (including Blackett Close) was set to a return period of 1000 years, to simulate a wall being built around it. Figure 10b shows that the validation area was then classified as unflooded.

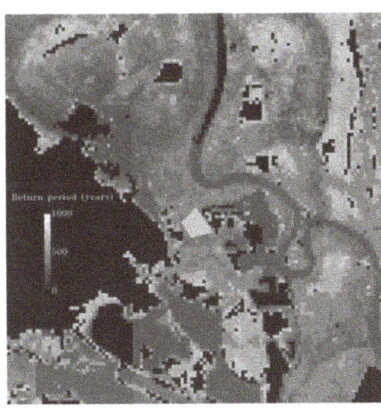
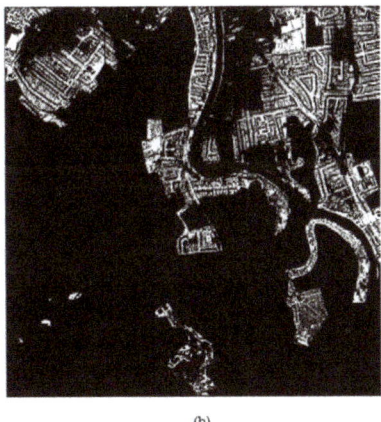

(a) (b)

Figure 10. (a) Return-period map for Staines with protective 'wall' around the validation area including Blackett Close (c/f Figure 6c, black areas masked out), and (b) urban flood classification using the SAR image of 13 February 2014, showing the unflooded validation area (white — flooded).

3.3. Case 3: Results Using Rural SAR Data, Precomputed FRP Maps, and Dynamic Flood Foresight Model Flood Extents

In this case the SAR waterline height map and the model waterline height map derived from the Flood Foresight modelled flood extent were combined, and the resulting threshold map was compared to the effective height map derived from the FRP maps (see Figure 4). The domains were again divided as for Case 1.

The Flood Foresight model was driven in its monitoring mode using telemetered streamflows from EA river gauges. Flood extents were estimated over the Staines and Tewkesbury domains every 3 h during the period 21 February 2014–29 February 2014. The flooding at Wraysbury was not modelled because insufficient gauge data were available there during the event.

For Staines, Figure 11 shows an example of the Flood Foresight flood extents both early (12 February 2014 18:00) and late (21 February 2014 18:00) in the flood centred on Blackett Close in Staines. The waterline height threshold maps were combined using the measured values for the SAR WLO standard deviation of 0.3 m and model WLO standard deviation of 0.4 m. For the model, the flood return period for both 13 February 2014 and 14 February 2014 corresponded to 1-in-20 years. The results for Blackett Close are shown in Table 4. The flood detection accuracies were similar to those for Cases 1 and 2, though for 13 February 2014, the false positive rate was slightly lower, leading to higher precision.

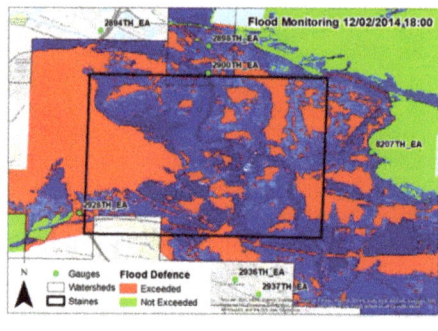 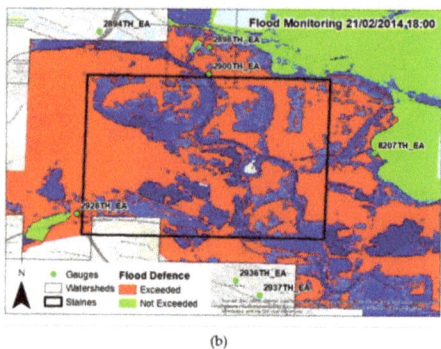

(a) (b)

Figure 11. Flood Foresight monitoring output centred on Blackett Close Staines (**a**) at 12 February 2014 18:00 near flood peak and (**b**) at 21 February 2014 18:00 when waters have receded. The blue shades indicate flooding to various depths, with darker blue being deeper flooding.

Table 4. Urban flood detection accuracy using rural SAR WLOs, FRP maps, and dynamic Flood Foresight model output (figures in brackets are for model output only).

Image + Model Output	Flood Detection Rate (Recall) (%)	Precision (%)	Critical Success Index (CSI) (%)
Blackett 13 February 2014 image + 13 February 2014 18:00 timestep model extent	100 (93)	91 (100)	91 (93)
Blackett 14 February 2014 image + 14 February 14 18:00 timestep model extent	98 (93)	100 (100)	98 (93)
Tewkesbury 25 July 2007 image + model maximum extent	74 (38)	90 (97)	69 (38)

It is also worth assessing how much the SAR data helped to improve the Blackett Close classification accuracies compared to using model data alone. This test was carried out by omitting the SAR data when the waterline height threshold maps were combined, so that the threshold was determined solely by the Flood Foresight model output. The resulting accuracies are shown in brackets in Table 4. Averaging results from both dates, there was a slight fall in urban flood detection rate but also a rise in precision. It seems that, in this case where the model was being driven by sufficient gauge data, the effect of including the rural SAR data was fairly marginal.

For Tewkesbury, Figure 12a shows the correspondence between the Flood Foresight maximum flood extent (modified by SAR data in the urban areas) and aerial photo flood extent, in urban and adjacent rural areas of Tewkesbury. The maximum extent represents the total extent of flooding throughout the flooding incident. The waterline height threshold maps were combined using the same weightings used in the Staines case. Urban flood detection results are shown in Table 4. The flood detection accuracy (74%) was slightly lower than the 88% achieved for Cases 1 and 2. An assessment was again made of how much the SAR data helped to improve the classification accuracy compared to using model data alone, by omitting the SAR data when the waterline height threshold maps were combined. The resulting flood detection accuracy using only model data (shown in brackets in Table 4) was 38%. This means that, in this case, the SAR WLOs provided a significant improvement compared to using model data alone.

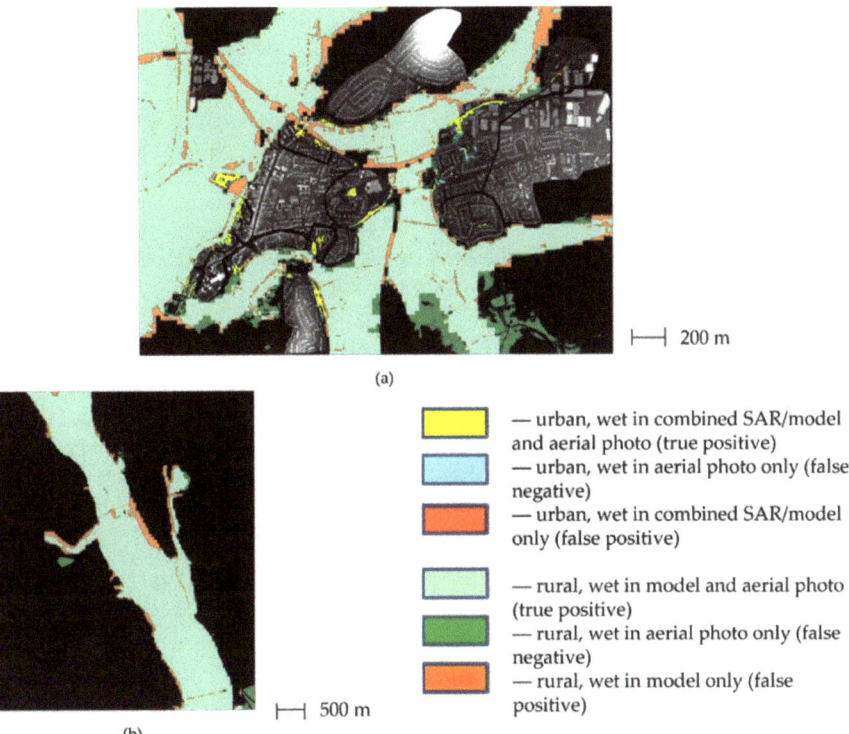

Figure 12. (a) Correspondence between the Flood Foresight model flood extent (combined with that of the SAR in urban areas) and aerial photo flood extent, in urban and rural areas of Tewkesbury, superimposed on the LiDAR image of the urban area (lighter grey – higher), and (b) correspondence between the Flood Foresight model flood extent and aerial photo flood extent in the long 6 km rural reach of the Severn immediately north-west of Tewkesbury (Mythe Bridge in the south of (b) is visible in the north-west of (a)).

The underestimation of urban flooding using only model data is likely to be because substantial surface water flooding was present in Tewkesbury during the event. This would have been present in the rural SAR data but not in the model output, as this predicts fluvial flooding only. In this event a good deal of rain fell in the vicinity of Tewkesbury itself, as well in the upper reaches of the Severn and Avon catchments, and there were extremely high flows in small local responsive catchments, such as the Isbourne just 20 km north-east of Tewkesbury [54]. In addition, Flood Foresight is a national-scale model being tested in the localized case of Tewkesbury that experienced a very small area of urban flooding. As a river model, it will mostly show water on the floodplain and is less likely to include urban flooding in its output. That Flood Foresight has modelled the rural flooding well in this case is shown in Figure 12b. This shows the correspondence between the modelled flood extent and the aerial photo flood extent in the long 6 km rural reach of the Severn immediately north-west of Tewkesbury. Over this reach, the hit rate was 96%, the precision 88%, and the CSI 85%. Figure 12a also shows that Flood Foresight predicted flooding well in the smaller rural areas immediately surrounding the urban area.

In dynamic situations such as the above, it is likely to be advantageous to use assimilation to combine the SAR and model WLO maps and assimilate a sequence of SAR WLO maps into the model WLO maps as they evolve over time, in order to improve the latter. As discussed previously, this would require overcoming the difficulty of updating the model WLO maps with height innovations, because of the simulation library approach used in the modelling.

4. Discussion

There is clear potential to integrate SAR data into Flood Foresight to improve the accuracy of the near real-time flood estimates produced by the model. The three examples described in this paper were each very small domains and used VHR SAR images, thus resulting in high levels of accuracy. However, VHR SAR data are rarely open access, and the reliance on LiDAR limits use of the method to urban regions of the globe that have been mapped by airborne LiDAR. It would obviously be more attractive to use global data sets that are easily accessible in an emergency. A suitable SAR candidate is Sentinel-1 because of its open access, the increased coverage of floods it provides globally using its wider swath, its high-resolution and preplanned acquisition capability, and the availability of georeferenced images in near real-time. Although perhaps not as accurate as using VHR SAR and LiDAR, the change detection method of Mason et al. (2021) [48] using Sentinel-1 and the WorldDEM DSM is capable of detecting urban flooding. Future work will investigate the merging of Sentinel-1 flood extents with model-derived flood hazard maps. If the approach can be applied to Sentinel-1 imagery, an automated process for a merged model-observational approach with coverage across much larger swathes of the country would be achievable and a cost-effective option for use by flood-response organisations.

Intuitively, merging modelled and remote observations of flooding has numerous benefits, some evidenced in the cases presented here. Although this study has shown an increase in accuracy and decreases in false negatives and false positives by combining SAR and model data, it is clear that there are high degrees of accuracy in the model that can provide valuable information to responders. The model has several advantages over the SAR due to its ability to provide near-real-time data 24/7/365 across the whole country. In addition, the locations in this study, although classed as 'urban', cannot be said to have a high degree of urban density. Towns and cities with higher density will increasingly suffer from shadow and layover issues affecting the ability to discriminate flood water in built-up areas. This will be exacerbated in instances where the coarser resolution Sentinel-1 data are used. In these cases, modelled inundation maps (such as those from Flood Foresight) are able to estimate flooded regions in the blind spots of the SAR. Expansion of the test areas to denser urban areas and across larger domains should therefore be explored further.

This study has explored merging model data generated from telemetered streamflow data from in-situ river gauges. Integrating SAR data (when available) was shown to improve the accuracy of flood detection in all three cases shown (Table 4). However, there is also the potential to improve the forecasting performance of the model using data assimilation, as alluded to in Section 2.3.2 although this will pose challenges given the precomputed simulation library approach used in Flood Foresight.

5. Conclusions

A number of conclusions can be drawn from the above results.

(1) Simply by using the rural SAR WLOs alone as in Case 1, a high urban flood detection accuracy (94%) and low false positive rate (9%) were achieved. However, this simple method cannot prevent urban areas that are low but defended from flooding from being detected as flooded.

(2) The Case 1 method using SAR backscatter only in rural areas was compared to the method of [42], which used SAR returns in both the rural and urban areas to detect urban flooding. It was found that the present method using SAR solely in rural areas was more accurate than that of [[42].

(3) The Case 2 method using rural SAR data and precomputed FRP maps should in theory have an advantage over the simple Case 1, as the flood return-period maps may contain information on defended regions in the urban area. There were no examples of this in the test areas, and consequently, the results for Case 1 and Case 2 were very similar. However, the potential advantage of using FRP maps was illustrated by simulating a defended region. In addition, the high accuracy obtained using the Case 2 method confirmed the findings of Tanguy et al. (2017) [43], who merged FRP maps with flood inundation maps derived from RADARSAT-2.

(4) Where the dynamic Flood Foresight model flood extents were combined with the rural SAR and FRP data (Case 3), then, for the Tewkesbury example, the rural SAR WLOs were able to provide a significant improvement compared to using model data alone, because there was significant surface-water flooding that was not reflected in the fluvially modelled flood extents. For the Blackett Close example, the classification improvement achieved by combining the rural SAR WLOs with the Flood Foresight model output was fairly marginal. However, it is interesting that, for these two examples, the results were almost no worse (indeed, for Tewkesbury, rather better), than if no dynamic model flood extents were used and the urban flood extent was predicted simply using rural SAR data and precomputed FRP maps (Case 2).

In summary, given the availability of VHR SAR and DSM data, the urban flood detection method merging model-derived FRP maps of the urban area with SAR returns in adjacent rural areas gave a high detection accuracy and was more accurate than that using SAR returns in both rural and urban areas. It would allow urban flood extents to be obtained in near real-time, and these could be used for emergency flood incident management and as observations for assimilation into flood forecasting models. The method could probably be extended to work with high resolution Sentinel-1 data, though further work is needed to confirm this. For the dynamical Flood Foresight model outputs, it is likely to be advantageous to use assimilation to combine the SAR and model WLO maps.

Author Contributions: Conceptualization, D.C.M. and J.B.; Methodology, D.C.M. and J.B.; Software, D.C.M., B.R.-R. and R.S.; Validation, D.C.M., B.R.-R. and R.S.; Formal Analysis, D.C.M. and J.B.; Investigation, D.C.M., B.R.-R. and R.S.; Writing—Original Draft Preparation, D.C.M. and J.B.; Writing—Review and Editing, D.C.M., J.B., S.L.D., S.V.-C. and H.L.C.; Supervision, S.L.D.; Project Administration, S.L.D.; Funding Acquisition, S.L.D. All authors have read and agreed to the published version of the manuscript.

Funding: This work was funded under the UK EPSRC grant EP/P002331/1 "Data Assimilation for the Resilient City (DARE)".

Acknowledgments: The authors are grateful to the UK Satellite Applications Catapult Centre for providing the CSK images under the CORSAIR Project. They are also grateful to the EA for provision of the LiDAR data.

Conflicts of Interest: The authors declare no conflict of interest.

References

1. Aerts, J.C.J.H.; Botzen, W.J.; Clarke, K.; Cutter, S.L.; Hall, J.W.; Merz, B.; Michel-Kerjan, E.; Mysiak, J.; Surminski, S.; Kunreuther, H. Integrating human behaviour dynamics into flood disaster risk assessment. *Nat. Clim. Chang.* **2018**, *8*, 193–199. [CrossRef]
2. Sharifi, A. Resilient urban forms: A macro-scale analysis. *Cities* **2019**, *85*, 1–14. [CrossRef]
3. García-Soriano, D.; Quesada-Román, A.; Zamorano-Orozco, J.J. Geomorphological hazards susceptibility in high-density urban areas: A case study of Mexico City. *J. South Am. Earth Sci.* **2020**, *102*, 102667. [CrossRef]
4. Quesada-Román, A.; Villalobos-Chacón, A. Flash flood impacts of Hurricane Otto and hydrometeorological risk mapping in Costa Rica. *Geogr. Tidsskr. J. Geogr.* **2020**, *120*, 142–155. [CrossRef]

5. Evans, E.P.; Ashley, R.; Hall, J.W.; Penning-Rowsell, E.C.; Saul, A.; Sayers, P.B.; Thorne, C.R.; Watkinson, A. *Foresight Flood and Coastal Defence Project: Scientific Summary*; Office of Science and Technology: London, UK, 2004.
6. Winsemius, H.C.; Aerts, J.C.J.H.; Van Beek, L.P.H.; Bierkens, M.F.P.; Bouwman, A.; Jongman, B.; Kwadijk, J.C.J.; Ligtvoet, W.; Lucas, P.L.; Van Vuuren, D.P.; et al. Global drivers of future river flood risk. *Nat. Clim. Chang.* **2016**, *6*, 381–385. [CrossRef]
7. Willner, S.N.; Otto, C.; Levermann, A. Global economic response to river floods. *Nat. Clim. Chang.* **2018**, *8*, 594–598. [CrossRef]
8. Li, Y.; Martinis, S.; Wieland, M.; Schlaffer, S.; Natsuaki, R. Urban Flood Mapping Using SAR Intensity and Interferometric Coherence via Bayesian Network Fusion. *Remote Sens.* **2019**, *11*, 2231. [CrossRef]
9. ICEYE. SAR Satellite Data Provider. Available online: https://www.iceye.com/sar-data/constellation-capabilities (accessed on 1 June 2021).
10. Pitt, M. Learning Lessons from the 2007 Floods. UK Cabinet Office Report. June 2008. Available online: http://archive.cabinetoffice.gov.uk/pittreview/thepittreview.html (accessed on 1 June 2021).
11. Brown, K.M.; Hambridge, C.H.; Brownett, J.M. Progress in operational flood mapping using satellite SAR and airborne LiDAR data. *Prog. Phys. Geog. Earth Environ.* **2016**, *40*, 186–214. [CrossRef]
12. Grimaldi, S.; Li, Y.; Pauwels, V.; Walker, J.P. Remote Sensing-Derived Water Extent and Level to Constrain Hydraulic Flood Forecasting Models: Opportunities and Challenges. *Surv. Geophys.* **2016**, *37*, 977–1034. [CrossRef]
13. García-Pintado, J.; Neal, J.C.; Mason, D.C.; Dance, S.L.; Bates, P.D. Scheduling satellite-based SAR acquisition for sequential assimilation of water level observations into flood modelling. *J. Hydrol.* **2013**, *495*, 252–266. [CrossRef]
14. García-Pintado, J.; Mason, D.C.; Dance, S.L.; Cloke, H.L.; Neal, J.C.; Freer, J.; Bates, P.D. Satellite-supported flood forecasting in river networks: A real case study. *J. Hydrol.* **2015**, *523*, 706–724. [CrossRef]
15. Mason, D.; Schumann, G.-P.; Neal, J.; Garcia-Pintado, J.; Bates, P. Automatic near real-time selection of flood water levels from high resolution Synthetic Aperture Radar images for assimilation into hydraulic models: A case study. *Remote Sens. Environ.* **2012**, *124*, 705–716. [CrossRef]
16. Giustarini, L.; Hostache, R.; Kavetski, D.; Chini, M.; Corato, G.; Schlaffer, S.; Matgen, P. Probabilistic Flood Mapping Using Synthetic Aperture Radar Data. *IEEE Trans. Geosci. Remote Sens.* **2016**, *54*, 6958–6969. [CrossRef]
17. Hostache, R.; Chini, M.; Giustarini, L.; Neal, J.; Kavetski, D.; Wood, M.; Corato, G.; Pelich, R.-M.; Matgen, P. Near-Real-Time Assimilation of SAR-Derived Flood Maps for Improving Flood Forecasts. *Water Resour. Res.* **2018**, *54*, 5516–5535. [CrossRef]
18. Cooper, E.S.; Dance, S.L.; García-Pintado, J.; Nichols, N.K.; Smith, P.J. Observation operators for assimilation of satellite observations in fluvial inundation forecasting. *Hydrol. Earth Syst. Sci.* **2019**, *23*, 2541–2559. [CrossRef]
19. Cooper, E.; Dance, S.; García-Pintado, J.; Nichols, N.; Smith, P. Observation impact, domain length and parameter estimation in data assimilation for flood forecasting. *Environ. Model. Softw.* **2018**, *104*, 199–214. [CrossRef]
20. Martinis, S.; Twele, A.; Voigt, S. Towards operational near real-time flood detection using a split-based automatic thresholding procedure on high resolution TerraSAR-X data. *Nat. Hazards Earth Syst. Sci.* **2009**, *9*, 303–314. [CrossRef]
21. Martinis, S.; Twele, A.; Voigt, S. Unsupervised Extraction of Flood-Induced Backscatter Changes in SAR Data Using Markov Image Modeling on Irregular Graphs. *IEEE Trans. Geosci. Remote Sens.* **2011**, *49*, 251–263. [CrossRef]
22. Martinis, S.; Kersten, J.; Twele, A. A fully automated TerraSAR-X based flood service. *ISPRS J. Photogramm. Remote Sens.* **2015**, *104*, 203–212. [CrossRef]
23. Pulvirenti, L.; Chini, M.; Pierdicca, N.; Guerriero, L.; Ferrazzoli, P. Flood monitoring using multi-temporal COSMO-SkyMed data: Image segmentation and signature interpretation. *Remote Sens. Environ.* **2011**, *115*, 990–1002. [CrossRef]
24. Pulvirenti, L.; Pierdicca, N.; Chini, M.; Guerriero, L. An algorithm for operational flood mapping from Synthetic Aperture Radar (SAR) data using fuzzy logic. *Nat. Hazards Earth Syst. Sci.* **2011**, *11*, 529–540. [CrossRef]
25. Twele, A.; Cao, W.; Plank, S.; Martinis, S. Sentinel-1-based flood mapping: A fully automated processing chain. *Int. J. Remote Sens.* **2016**, *37*, 2990–3004. [CrossRef]
26. D'Addabbo, A.; Refice, A.; Pasquariello, G.; Lovergine, F.P.; Capolongo, D.; Manfreda, S. A Bayesian Network for Flood Detection Combining SAR Imagery and Ancillary Data. *IEEE Trans. Geosci. Remote Sens.* **2016**, *54*, 3612–3625. [CrossRef]
27. D'Addabbo, A.; Refice, A.; Lovergine, F.P.; Pasquariello, G. DAFNE: A Matlab toolbox for Bayesian multi-source remote sensing and ancillary data fusion, with application to flood mapping. *Comput. Geosci.* **2018**, *112*, 64–75. [CrossRef]
28. Matgen, P.; Hostache, R.; Schumann, G.; Pfister, L.; Hoffmann, L.; Savenije, H. Towards an automated SAR-based flood monitoring system: Lessons learned from two case studies. *Phys. Chem. Earth Parts A/B/C* **2011**, *36*, 241–252. [CrossRef]
29. Giustarini, L.; Hostache, R.; Matgen, P.; Schumann, G.; Bates, P.D.; Mason, D.C. A change detection approach to flood mapping in urban areas using TerraSAR-X. *IEEE. Trans. Geosci. Remote Sens.* **2013**, *51*, 2417–2430. [CrossRef]
30. Pierdicca, N.; Pulvirenti, L.; Chini, M.; Guerriero, L.; Candela, L. Observing floods from space: Experience gained from COSMO-SkyMed observations. *Acta Astronaut.* **2013**, *84*, 122–133. [CrossRef]
31. Schumann, G.; di Baldassarre, G.D.; Bates, P.D. The utility of spaceborne radar to render flood inundation maps based on multialgorithm ensembles. *IEEE Trans. Geosci. Remote Sens.* **2009**, *47*, 2801–2807. [CrossRef]
32. Schlaffer, S.; Chini, M.; Giustarini, L.; Matgen, P. Probabilistic mapping of flood-induced backscatter changes in SAR time series. *Int. J. Appl. Earth Obs. Geoinf.* **2017**, *56*, 77–87. [CrossRef]
33. Westerhoff, R.S.; Kleuskens, M.P.H.; Winsemius, H.C.; Huizinga, H.J.; Brakenridge, G.R.; Bishop, C. Automated global water mapping based on wide-swath orbital synthetic-aperture radar. *Hydrol. Earth Syst. Sci.* **2013**, *17*, 651–663. [CrossRef]

34. Benoudjit, A.; Guida, R. A Novel Fully Automated Mapping of the Flood Extent on SAR Images Using a Supervised Classifier. *Remote Sens.* **2019**, *11*, 779. [CrossRef]
35. Nemni, E.; Bullock, J.; Belabbes, S.; Bromley, L. Fully Convolutional Neural Network for Rapid Flood Segmentation in Synthetic Aperture Radar Imagery. *Remote Sens.* **2020**, *12*, 2532. [CrossRef]
36. Ohki, M.; Yamamoto, K.; Tadono, T.; Yoshimura, K. Automated Processing for Flood Area Detection Using ALOS-2 and Hydrodynamic Simulation Data. *Remote Sens.* **2020**, *12*, 2709. [CrossRef]
37. Soergel, U.; Thoennessen, U.; Stilla, U. Visibility analysis of man-made objects in SAR images. In Proceedings of the 2003 2nd GRSS/ISPRS Joint Workshop on Remote Sensing and Data Fusion over Urban Areas, Berlin, Germany, 22–23 May 2003. [CrossRef]
38. Pulvirenti, L.; Chini, M.; Pierdicca, N.; Boni, G. Use of SAR Data for Detecting Floodwater in Urban and Agricultural Areas: The Role of the Interferometric Coherence. *IEEE Trans. Geosci. Remote Sens.* **2016**, *54*, 1532–1544. [CrossRef]
39. Mason, D.C.; Speck, R.; Devereux, B.; Schumann, G.J.-P.; Neal, J.C.; Bates, P.D. Flood detection in urban areas using TerraSAR-X. *IEEE. Trans. Geosci. Remote Sens.* **2010**, *48*, 882–894. [CrossRef]
40. Mason, D.C.; Davenport, I.; Neal, J.C.; Schumann, G.; Bates, P.D. Near Real-Time Flood Detection in Urban and Rural Areas Using High-Resolution Synthetic Aperture Radar Images. *IEEE Trans. Geosci. Remote Sens.* **2012**, *50*, 3041–3052. [CrossRef]
41. Mason, D.; Giustarini, L.; Garcia-Pintado, J.; Cloke, H. Detection of flooded urban areas in high resolution Synthetic Aperture Radar images using double scattering. *Int. J. Appl. Earth Obs. Geoinf.* **2014**, *28*, 150–159. [CrossRef]
42. Mason, D.C.; Dance, S.L.; Vetra-Carvalho, S.; Cloke, H.L. Robust algorithm for detecting floodwater in urban areas using synthetic aperture radar images. *J. Appl. Remote Sens.* **2018**, *12*, 045011. [CrossRef]
43. Tanguy, M.; Chokmani, K.; Bernier, M.; Poulin, J.; Raymond, S. River flood mapping in urban areas combining Radarsat-2 data and flood return period data. *Remote Sens. Environ.* **2017**, *198*, 442–459. [CrossRef]
44. Chini, M.; Pelich, R.-M.; Pulvirenti, L.; Pierdicca, N.; Hostache, R.; Matgen, P. Sentinel-1 InSAR Coherence to Detect Floodwater in Urban Areas: Houston and Hurricane Harvey as A Test Case. *Remote Sens.* **2019**, *11*, 107. [CrossRef]
45. Li, Y.; Martinis, S.; Wieland, M. Urban flood mapping with an active self-learning convolutional neural network based on TerraSAR-X intensity and interferometric coherence. *ISPRS J. Photogramm. Remote Sens.* **2019**, *152*, 178–191. [CrossRef]
46. Lin, Y.N.; Yun, S.-H.; Bhardwaj, A.; Hill, E.M. Urban Flood Detection with Sentinel-1 Multi-Temporal Synthetic Aperture Radar (SAR) Observations in a Bayesian Framework: A Case Study for Hurricane Matthew. *Remote Sens.* **2019**, *11*, 1778. [CrossRef]
47. Iervolino, P.; Guida, R.; Iodice, A.; Riccio, D. Flooding water depth estimation with high-resolution SAR. *IEEE Trans. GeoSci. Remote Sens.* **2015**, *53*, 2295–2307. [CrossRef]
48. Mason, D.C.; Dance, S.L.; Cloke, H.L. Floodwater detection in urban areas using Sentinel-1 and WorldDEM data. *J. Appl. Remote Sens.* **2021**, *15*, 032003. [CrossRef]
49. Wessel, B.; Huber, M.; Wohlfart, C.; Marschalk, U.; Kossmann, K.; Roth, A. Accuracy assessment of the global TanDEM-X Digital Elevation Model with GPS data. *ISPRS J. Photogramm. Remote Sens.* **2018**, *139*, 171–182. [CrossRef]
50. Revilla-Romero, B.; Shelton, K.; Wood, E.; Berry, R.; Bevington, J.; Hankin, B.; Lewis, G.; Gubbin, A.; Griffiths, S.; Barnard, P.; et al. Flood Foresight: A near-real time flood monitoring and forecasting tool for rapid and predictive flood impact assessment. *Geophys. Res. Abstr.* **2017**, *19*, 1230.
51. Bradbrook, K. JFLOW: A multiscale two-dimensional dynamic flood model. *Water Environ. J.* **2006**, *20*, 79–86. [CrossRef]
52. Environment Agency. Real-time Flood Impacts Mapping. 2019. Available online: https://assets.publishing.service.gov.uk/government/uploads/system/uploads/attachment_data/file/844094/Real-time_flood_impacts_mapping_-_report.pdf (accessed on 1 June 2021).
53. Thorne, C. Geographies of UK flooding in 2013/4. *Geogr. J.* **2014**, *180*, 297–309. [CrossRef]
54. Stuart-Menteth, A. *UK Summer 2007 Floods, 2007*; Risk Management Solutions: Newark, CA, USA, 2007.
55. Neal, J.C.; Keef, C.; Bates, P.D.; Beven, K.; Leedal, D. Probabilistic flood risk mapping including spatial dependence. *Hydrol. Process.* **2012**, *27*, 1349–1363. [CrossRef]
56. Brisco, B.; Touzi, R.; Van der Sanden, J.J.; Charbonneau, F.; Pultz, T.J.; D'Iorio, M. Water resource applications with RA-DARSAT-2—A preview. *Int. J. Digit. Earth* **2008**, *1*, 130–147. [CrossRef]
57. Marconcini, M.; Metz-Marconcini, A.; Üreyen, S.; Palacios-Lopez, D.; Hanke, W.; Bachofer, F.; Zeidler, J.; Esch, T.; Gorelick, N.; Kakarla, A.; et al. Outlining where humans live, the World Settlement Footprint 2015. *Sci. Data* **2020**, *7*, 1–14. [CrossRef] [PubMed]
58. Definiens, A.G. *Definiens Developer 8 User Guide, Document Version 1.2.0*; Definiens Documentation: Munich, Germany, 2012.
59. Aitken, A.C. IV.—On Least Squares and Linear Combination of Observations. *Proc. R. Soc. Edinb.* **1936**, *55*, 42–48. [CrossRef]

Article

Evaluation of CYGNSS Observations for Flood Detection and Mapping during Sistan and Baluchestan Torrential Rain in 2020

Mahmoud Rajabi *, Hossein Nahavandchi and Mostafa Hoseini

Department of Civil and Environmental Engineering, Norwegian University of Science and Technology NTNU, 7491 Trondheim, Norway; hossein.nahavandchi@ntnu.no (H.N.); mostafa.hoseini@ntnu.no (M.H.)
* Correspondence: mahmoud.rajabi@ntnu.no; Tel.: +47-92332254

Received: 9 June 2020; Accepted: 16 July 2020; Published: 18 July 2020

Abstract: Flood detection and produced maps play essential roles in policymaking, planning, and implementing flood management options. Remote sensing is commonly accepted as a maximum cost-effective technology to obtain detailed information over large areas of lands and oceans. We used remote sensing observations from Global Navigation Satellite System-Reflectometry (GNSS-R) to study the potential of this technique for the retrieval of flood maps over the regions affected by the recent flood in the southeastern part of Iran. The evaluation was made using spaceborne GNSS-R measurements over the Sistan and Baluchestan provinces during torrential rain in January 2020. This area has been at a high risk of flood in recent years and needs to be continuously monitored by means of timely observations. The main dataset was acquired from the level-1 data product of the Cyclone Global Navigation Satellite System (CYGNSS) spaceborne mission. The mission consisted of a constellation of eight microsatellites with GNSS-R sensors onboard to receive forward-scattered GNSS signals from the ocean and land. We first focused on data preparation and eliminating the outliers. Afterward, the reflectivity of the surface was calculated using the bistatic radar equations formula. The flooded areas were then detected based on the analysis of the derived reflectivity. Images from Moderate-Resolution Imaging Spectroradiometer (MODIS) were used for evaluation of the results. The analysis estimated the inundated area of approximately 19,644 km^2 (including Jaz-Murian depression) to be affected by the flood in the south and middle parts of the Sistan and Baluchestan province. Although the main mission of CYGNSS was to measure the ocean wind speed in hurricanes and tropical cyclones, we showed the capability of detecting floods in the study area. The sensitivity of the spaceborne GNSS-R observations, together with the relatively short revisit time, highlight the potential of this technique to be used in flood detection. Future GNSS-R missions capable of collecting the reflected signals from all available multi-GNSS constellations would offer even more detailed information from the flood-affected areas.

Keywords: CYGNSS; flood detection; Sistan and Baluchestan; flood mapping; GNSS-R

1. Introduction

Natural disasters are the reason for many serious disturbances to communities and the environment. There have been many human, environmental, social, and economic losses, which are beyond the power of the community to tolerate [1]. Floods have been considered as one of the most catastrophic events, causing extensive damage to the artificial and natural environment and devastation to human settlements [2]. Economic losses due to the effects of damaging floods have increased significantly around the world [3]. Flooding happens when water bodies overflow riversides, lakes, dams, or dikes in low-lying lands during heavy rainfall [4]. The higher temperature at the Earth's surface leads to increased evaporation and greater overall precipitation [5]. Increased precipitation, although associated with inland flooding, can also increase the risk of coastal flooding [6].

Flood detection, and subsequently, produced maps, are beneficial in two important phases: During the flood, when we need emergency management planning, and after the flood, for land use planning, defining construction standards, and damage assessment [7]. Heavy precipitation has led floods to occur more frequently in different countries, which have drawn considerable attention over the past years. There are many regions of Iran affected by floods, for instance, heavy rainfall from mid-March to April 2019 led to flooding in 28 of 31 provinces, with the most severe flooding occurring in Golestan, Fars, Khuzestan, and Lorestan [8]. The recent torrential rain in mid-January 2020 in the southeastern region of Iran caused a devastating flood in the Sistan and Baluchestan province. We investigated the latter case in this study.

Land surveying and airborne observations are the traditional methods for flood detection, but when when flood detection is conducted on a large scale, these methods are costly and slow. Space-based Remote Sensing (RS) can be considered as a practical alternative that provides up-to-date information from various sensors that have been onboard different satellites. However, there are some limitations in using RS data products for the study of flooding. For instance, optical RS can have its limitations during severe weather conditions and during night. Therefore, in some cases before and after a flood event, the optical RS imagery does not provide the required information [7]. Radar RS in the microwave spectrum can surpass these restrictions because the wave can penetrate clouds and vegetation and can effectively work at night. Among the several radars RS sensors currently in operation, Synthetic Aperture Radar (SAR) imagery provides high spatial resolution data which is typically based on a monostatic configuration. However, the revisit time of satellites with the configuration of the monostatic radar (single satellites), like SAR, is long (more than one week) and cannot offer the desirable continuous high temporal resolution for flood detection purposes. Accordingly, owing to the highly dynamic nature of the flood, SAR images are not used operationally during floods [9–11].

The primary services of the Global Navigation Satellite System (GNSS) are positioning, navigation, and timing. Besides, many other applications, including GNSS RS, have been introduced in recent decades. Measurements made by GNSS RS techniques provide valuable information about different components of the Earth system. Observations of the GNSS signals passing through the atmosphere have been employed to study the atmospheric layers and their variabilities [12,13]. GNSS signals after reflection from the Earth's surface can also provide information about the reflecting surface. These reflections have been used to study various parameters of the Earth's surface and water cycles, such as snow depth [14], ice height and sea level [15], soil moisture [16], vegetation [17], flood [11,18], ocean eddies [19], wind speed [20], salinity [21], etc.

Global Navigation Satellite System Reflectometry (GNSS-R) is an innovative technique aimed at deriving geophysical parameters by analyzing GNSS signals reflected off the Earth's surface in a bistatic geometry. This technique is an efficient microwave remote sensing approach that utilizes transmitted navigation signals as sources of opportunity. Numerous GNSS satellites, including GPS, Galileo, GLONASS, and Bei-Dou/Compass, are currently transmitting navigation signals based on spread-spectrum technology. Thus, a constellation of GNSS-R small satellites, at a lower cost compared to ordinary RS satellites, can provide a much shorter revisit time using low-cost, low-power passive sensors. Many earlier studies have introduced the applications of GNSS-R on the oceans, land, and ice [22–24].

The soil moisture, surface roughness, vegetation, and topography are parameters which affect microwave signals. GNSS-R signals as a bistatic radar are also affected by those parameters [25]. However, GNSS signals are at the L-band, which is ideal for soil moisture and surface water remote sensing due to the higher capacity to penetrate vegetation compared to shorter wavelengths [16]. In addition, this technique uses the bistatic configuration, which has a lower sensitivity to surface roughness relative to monostatic [26]. The signals reflected off the surface have a direct relation with surface water and moisture content [11]. For example, the rise of soil moisture leads to increase the signal strength. Using this mechanism could contribute to detecting soil saturation, flooded area, and inland water.

The Cyclone Global Navigation Satellite System (CYGNSS) mission is a constellation of eight microsatellites, each with a GNSS-R receiver onboard. The receiver can track and process four GPS signals simultaneously. The tracked GPS L1 C/A signals after reflection from the Earth's surface are used to produce Delay Doppler Maps (DDMs). The overall median revisit time is 2.8 h, and the mean revisit time is 7.2 h [27]. Theoretically, the footprint of reflection received by CYGNSS is nearly 0.5 km × 0.5 km. For the ocean, which has a very rough surface, the spatial resolution is approximately 25 km × 25 km [28,29]. Table 1 shows CYGNSS microsatellite parameters retrieved from [16,23]. The main mission of CYGNSS is to measure the ocean surface wind speed in hurricanes and tropical cyclones, so a relatively low orbital inclination was designed for the satellites. CYGNSS continuously makes measurements over the oceans and provides useful information over the land [29]. CYGNSS offers distinct features compared to other remote sensing techniques such as optical and active monostatic radar. It uses a passive sensor at the L-band frequency wave, which works in all weather conditions regardless of the time of the day, i.e., it can penetrate clouds, fog, rain, storms, and vegetation, and works at night, unlike optical sensors. The CYGNSS constellation of eight microsatellites provides a relatively short revisit time with global coverage over equatorial regions. The products of CYGNSS are publicly available over the oceans and land.

Table 1. The Cyclone Global Navigation Satellite System (CYGNSS) satellite parameters.

Parameters	Description
Orbit	LEO, ~520 km, Nonsynchronous
Period	95.1 min
Spatial Resolution	~25 km × 25 km (incoherent), ~0.5 km × 5 km (coherent, theoretical)
Revisit Times	2.8 h median, 7.2 h mean
Polarization of the reflectometry antennas	LHCP
Coverage	−38 < Latitude < 38 & −180 < Longitude < 180
Type of Data which is relevant	Observe GPS L1 C/A signals and Delay Doppler Maps

Radar remote sensing for soil moisture retrieval and surface water detection is common using both monostatic [7,30] and bistatic geometry. The sensitivity of spaceborne GNSS-R (as a bistatic radar) to surface water and soil moisture has been widely studied [11,16,18,23,31–33]. Most of the studies have used observations from ground-based or space-based receivers, e.g., CYGNSS or Technology Demonstration Satellite-1 (TDS-1). Observational evidence demonstrates that GNSS-R is highly sensitive to inland surface waters, e.g., lakes and rivers [34].

Sistan and Baluchestan is one of the driest regions of Iran, with a slight increase in rainfall from east to west, and is a province at a high risk of flooding. The aim of this study was to indicate the capability of spaceborne GNSS-R for detecting and mapping of flood in the south part of Iran. The methodology for preparing and processing data is the same as those used described by the authors of [16,18].

2. Study Area

The Sistan and Baluchestan province is located in the east and southeast of Iran (58°55′–63°20′ E longitude and 25°04′–31°25′ N latitude), bordering Pakistan and Afghanistan, and its capital is Zahedan. This province is the second largest province in Iran with an area of 180,726 km^2 and a population of about 2.5 million. Figure 1A shows the location of this province on the Maphill Earth map. There is a depression in the study area known as the Hamun-Jaz-Murian basin, which is part of the central plateau basin. This basin is located in the southeast of Iran between 56°17′ and 61°25′ E longitude and 26°32′ and 29°35′ N latitude (Figure 1B). Its total area is about 69,390 km^2, of which 44% is mountains. The depression belongs to Kerman and Sistan-Baluchistan provinces [35]. Figure 1C shows a flooded region in IranShahr, which is one of the cities in this province.

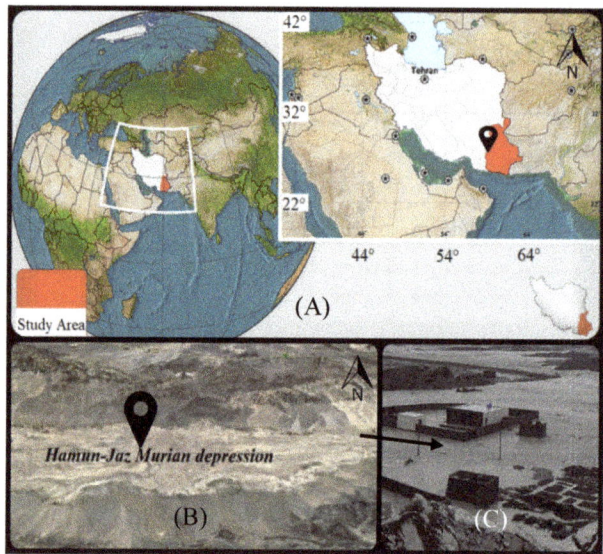

Figure 1. (**A**) The red region shows the location of the Sistan and Baluchestan province in the southeast of Iran, which is our study area. (**B**) The Hamun Jaz-Murian depression, which is located between the Kerman and Sistan and Baluchestan provinces. (**C**) IranShahr, one of the flooded cities in the study area.

Sistan and Baluchestan is one of the warmest regions in Iran, with a desert climate and an average daily temperature of 29 degrees centigrade. For several months of the year, it is warm at temperatures continuously above 25 degrees centigrade, and temperatures sometimes exceed above 40 degrees centigrade. Figure 2 illustrates the average precipitation per day over 20 years. As can be seen, 0.40 mm/day rainfall is normal during January in the province, but between 10 January and 12 January 2020, this amount is over 100 mm. Figure 3 shows the precipitation rate from 8 January to 13 January 2020.

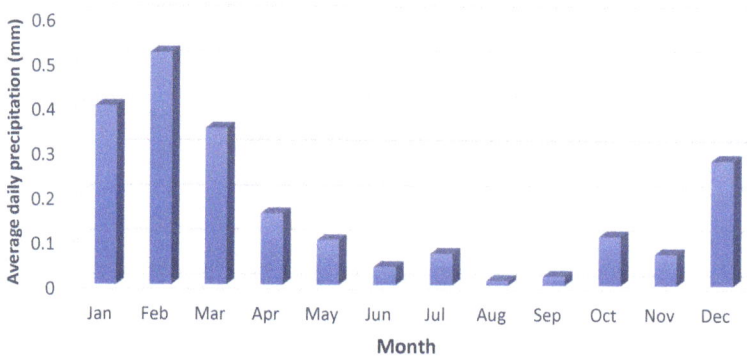

Figure 2. Average daily precipitation data collected from three meteorological stations in the Sistan and Baluchestan province based on the average values of the last 20 years [36].

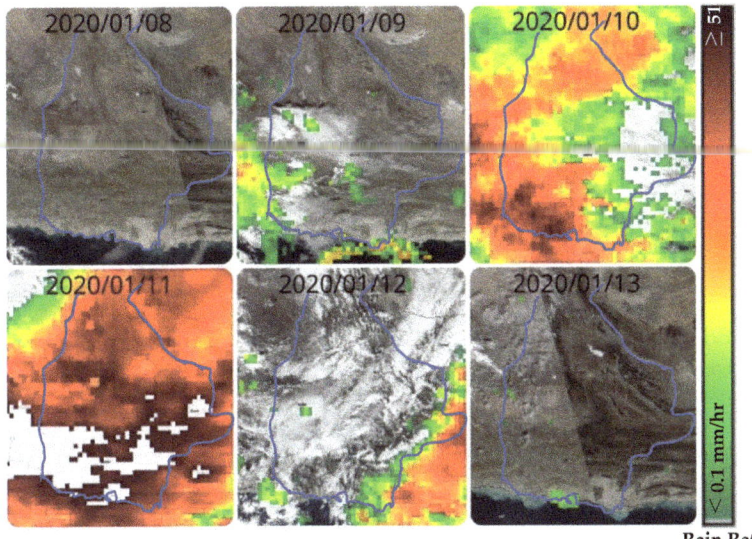

Figure 3. Rate of torrential precipitation in the Sistan and Baluchestan province over the period of six days from 8 January to 13 January, 2020. The maps were generated using the data provided by the authors of [37].

3. Data Set Description

3.1. CYGNSS data

There are three levels of CYGNSS data products in version 2.1, which represent the second post-provisional based on calibrated and validated level 1 algorithms. The level 1 (L1) dataset contains the measurement of surface Normalized Bistatic Radar Cross Section (NBRCS). The level 2 (L2) dataset includes derived ocean surface wind speed and Mean Square Slope (MSS). The level 3 (L3) dataset delivers hourly averaged wind speed and MSS on a 0.2 degree × 0.2 degree grid.

We used CYGNSS L1 data as the lowest level of the available data products. The format of the data is NetCDF (Network Common Data Form). Daily observations of each of the eight CYGNSS satellites are included in a NetCDF file. Accordingly, there are up to eight files for every Day Of a Year (DOY). The daily base data is available free of charge on the website of Physical Oceanography Distributed Active Archive Center (PO. DAAC) of NASA's Jet Propulsion Laboratory (JPL) at https://podaac.jpl.nasa.gov. Table 2 shows the main variables of the L1 data [24] which were used in this study.

Table 2. The Cyclone Global Navigation Satellite System (CYGNSS) data source parameters.

Parameters	Description
ddm_snr	Delay Doppler Map (DDM) signal-to-noise ratio, in dB
gps_tx_power_db_w	GPS transmit power, in dB.
rx_to_sp_range	Distance between the CYGNSS spacecraft and the specular point, in meters.
tx_to_sp_range	Distance between the GPS spacecraft and the specular point, in meters.
gps_ant_gain_db_i	GPS transmit antenna gain. Antenna gain in the direction of the specular point, in dBi
sp_rx_gain	Specular point Rx antenna gain. The receive antenna gains in the direction of the specular point, in decibel isotropic (dBi).
quality_flags	Per-DDM quality flags
sp_lat	Specular point latitude, in degrees North
sp_lon	Specular point longitude, in degrees East
sp_inc_angle	The specular point incidence angle, in degrees

3.2. Satellite Image

Moderate-Resolution Imaging Spectroradiometer (MODIS) is an advanced sensor on the Terra and Aqua Spacecraft for gathering data through a broad spectrum of electromagnetic waves. Terra was the first satellite of the Earth Observing System (EOS) program and was launched on 18 December, 1999. It passes north to south over the equator in the morning. Aqua is the second EOS satellite which carries a MODIS sensor and passes south to north across the equator in the afternoon. Terra and Aqua MODIS cover the Earth's surface every one to two days. The sensors onboard these satellites measure 36 spectral bands from 0.405 µm to 14.385 µm. The data is released by different resolutions, i.e., 250 m (bands 1–2), 500 m (bands 3–7), and 1000 m (bands 8–36). The MODIS data is accessible at https://modis.gsfc.nasa.gov and can be used for a significant number of applications in the land, atmosphere, and, ocean [38]. Figure 4 shows the false-color images of the Sistan and Baluchestan province (also regions of the Kerman and Hormozgan provinces) before the flood (A) and during the flood (B). The images were acquired by MODIS (bands 7–2–1) on 8 January and 13 January 2020. These images were used here for validation purposes.

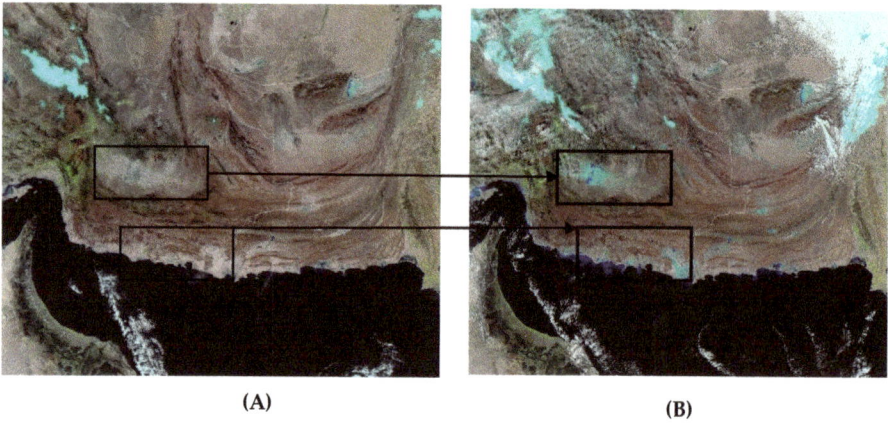

(A)   (B)

Figure 4. Moderate-Resolution Imaging Spectroradiometer (MODIS) images of the study area (**A**) before the flooding on 8 January 2020, and (**B**) during the flood on 13 January 2020. The dark blue regions are the inundated areas. The clouds in the image are shown with light blue which can be distinguished from the inundated areas [39].

4. Method and Discussion

The methodology in the current paper includes five main steps, as illustrated in Figure 5. The steps are: (1) Data collection, (2) data preparation, (3) calculating the surface reflectivity, (4) data calibration, and (5) flood detection and validation. Each step is described as follows.

4.1. The Bistatic Radar Equations

Radar is a system for detecting targets and deriving information such as position, velocity, and reflectivity signature from the detected objects [40]. It transmits a signal and receives the echo after it is reflected by a target. The types of radar systems based on the location of the transmitter (TX) and the receiver (RX) can be divided into colocated or monostatic radars, which measure backscattered signals, and separated or bistatic radars, which measure forward-scattered signals. The main difference between monostatic and bistatic radars is the separation of the transmitter and receiver [41]. Figure 6 shows monostatic and bistatic constellation for satellites.

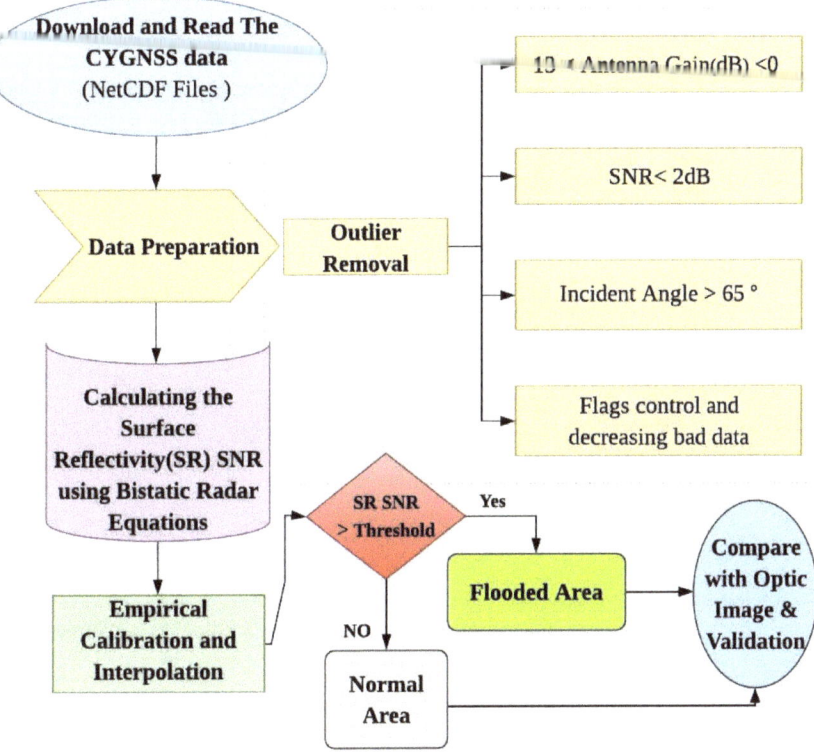

Figure 5. Methodology flowchart based on the bistatic radar concept and using CYGNSS data.

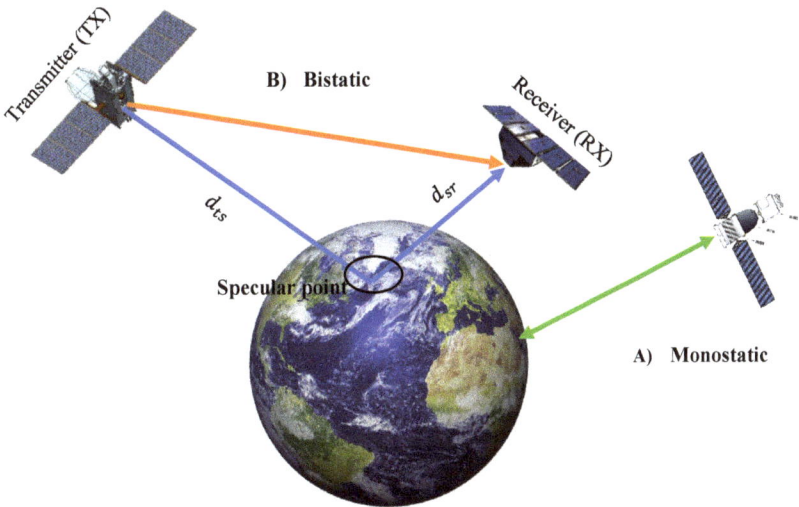

Figure 6. Simple schematic view of mono (**A**) and bistatic (**B**) radars constellation for satellite remote sensing.

The CYGNSS and GPS constellations form a bistatic radar system. The GPS satellites transmit circularly polarized microwave signals which are collected by the CYGNSS reflectometry receivers after forward-scattering from the Earth's surface. The scattered signals contain valuable information about the physical properties of the reflecting surface. Inland waters can be detected by CYGNSS, assuming a coherent forward-scattering mechanism [33,34,42]. The peak value of coherent scattered power is defined as [22,23,43]:

$$P_{RL}^{coh} = \frac{P_R^t \, G^t \, G^r}{(d_{ts} + d_{sr})^2} \left(\frac{\lambda}{4\pi}\right)^2 \Gamma_{RL} \tag{1}$$

where P_{RL}^{coh} is the peak value of coherently received power, R denotes the right-handed circular polarization (RHCP) GPS transmit antenna, and L is related to the left-handed circular polarization (LHCP) of forward-scattered signals collected by the downward-looking antenna. P_R^t is the transmitted power, G^t is the gain of the transmitter antenna, G^r is the gain of the receiver antenna, λ is the GPS L1 wavelength (~0.19 m), and d_{ts} is the distance between the specular reflection point and the GPS transmitter, while d_{sr} is the distance between the specular reflection point and the GNSS-R receiver and Γ_{RL} is the surface reflectivity along with the incidence angle. In addition to the mentioned parameters P_{RL}^{coh} is affected by system noise. Therefore, signal-to-noise ratio (SNR) could be defined as:

$$SNR = \frac{P_{RL}^{coh}}{N} = \frac{P_R^t \, G^t \, G^r}{(d_{ts} + d_{sr})^2} \left(\frac{\lambda}{4\pi}\right)^2 \frac{\Gamma_{RL}}{N} \tag{2}$$

where N is the noise value. Since the magnitude of the SNR is not equal to the reflected power, the surface reflectivity or corrected SNR along with the incidence angle could be computed using:

$$SNR_c = \frac{\Gamma_{RL}}{N} = SNR \frac{(d_{ts} + d_{sr})^2}{P_R^t G^t \, G^r} \left(\frac{4\pi}{\lambda}\right)^2 \tag{3}$$

Finally, the SNR_c in decibel (dB) is:

$$SNR_{c\,dB} = SNR_{dB} + 10 \log(\frac{(d_{ts} + d_{sr})^2}{P_R^t G^t \, G^r} \left(\frac{4\pi}{\lambda}\right)^2) \tag{4}$$

This parameter ($SNR_{c\,dB}$) is strongly related to the hydrological conditions of the land surface [18,34]. In this study, the following CYGNSS L1 variables were used for the calculation of the surface reflectivity:

- ddm_snr ($SNR_{dB} = 10\log(S_{max}/N_{avg})$ with S_{max} being the maximum value in a single DDM bin and N_{avg} is the average raw noise counts per-bin
- gps_tx_power_db_w (P_R^t)
- gps_ant_gain_db_i (G^t)
- sp_rx_gain (G^r)
- rx_to_sp_range (d_{sr})
- tx_to_sp_range (d_{ts})

The parameter λ is the wavelength of the GPS L1 carrier (~0.19 m). We converted all the values to the dB scale (some of them were already in dB within the CYGNSS files).

4.2. Data Preparation and Calibration

Before and after using Equation (4), we employed several corrections and data editions and outlier identification as follows:

- GPS transmitter bias: GPS transmit powers are approximate estimates with some biases which should be considered. The main sources of these biases could be unknown transmitting powers of GPS satellites and the biases in P_R^t associated with GPS pseudorandom noise (PRN) codes [16,44]. We used empirical calibration developed by Chew et al. (2018) for CYGNSS products. Table 3 shows the magnitude of the biases which should be corrected during the estimation of $SNR_{c\ dB}$ [15].
- Incidence angle: This parameter also affects a coherent reflection when the incidence angles are above 40 degrees or 50 degrees and was negligible for our purpose [34], but we deleted data with an incidence angle of more than 65 degrees.
- Quality Control Flags: The Level 1A data product used in this study was refined by applying a set of quality control flags designed and included in the data to indicate potential problems [27,45]. The specific flags we used were 2, 4, 5, 8, 16, and 17, which were related to S-band transmitter powered up, spacecraft attitude error, black body DDM, DDM is a test pattern, the direct signal in DDM, and low confidence in the GPS EIR estimate, respectively. Based on the work by Chew et al. (2018) on soil moisture, we removed data with those quality flags in this study.
- Additional correction and removal: We removed data with SNR_{dB} less than 2dB and CYGNSS antenna gain of less than 0 dB or more than 13 dB. These corrections were empirical and are not standardized, but have been shown to be beneficial [16].

Table 3. Empirical biases in $SNR_{c\ dB}$ according to pseudorandom noise (PRN).

PRN	Bias (dB)	PRN	Bias (dB)	PRN	Bias (dB)	PRN	Bias (dB)
1	1.017	9	1.498	17	0.256	25	0.880
2	0.004	10	−0.783	18	−0.206	26	0.163
3	1.636	11	−0.230	19	−0.206	27	0.409
4	-	12	−1.021	20	0.345	28	−0.712
5	−0.610	13	0.007	21	−0.909	29	−1.032
6	0.24	14	−0.730	22	−0.838	30	0.877
7	−0.709	15	−0.376	23	−0.858	31	−0.562
8	0.605	16	−0.481	24	1.140	32	−0.819

Figure 7 shows the statistical information for corrected SNR using three days of CYGNSS data during the flood time. Figure 7A shows the calculated surface reflectivity SNR of CYGNSS tracks before (left side) and after (right side) the data preparation. As can be seen in the middle part of the figure, some of the measurements that may be misleading were removed. Figure 7B,C show the distribution of the measurements with respect to the incidence angle and antenna gain. Despite the fact that the data rectification procedure discarded about 48% of the observations, CYGNSS still provided enough data to detect the flood. The flooding period continued until 17 January 2020. We analyzed a dataset consisting of three days of CYGNSS observations to reduce the effect of losing a significant portion of the data.

4.3. Interpolation

An interpolation process was used here to retrieve a representative grid from the CYGNSS observation points. As shown in Figure 7A, the data derived from CYGNSS have irregular structures based on the satellite along-tracks. We used the natural neighbor interpolation method for gridding. The method was developed by Sibson [46] and is a multivariate interpolation according to Voronoi tessellation [47]. The principal formula is [48]:

$$G(x,y) = \sum_{i=1}^{N} w_i\, f(x_i, y_i) \quad (5)$$

where G is the estimated value at (x,y), $w_i = Q_k/R_k$ is the weights, and $f(x_i, y_i)$ is the known data at (x_i, y_i), R_k is the area of the initial Voronoi diagram element for point $P_k = (x_i, y_i)$. Q_k is the intersection area of R_k and newly constructed element for the point (x,y). Therefore, the method algorithm is the algorithm to insert an additional point into the existing Voronoi diagram. Figure 8 illustrates the visual view of the natural neighbor interpolation method.

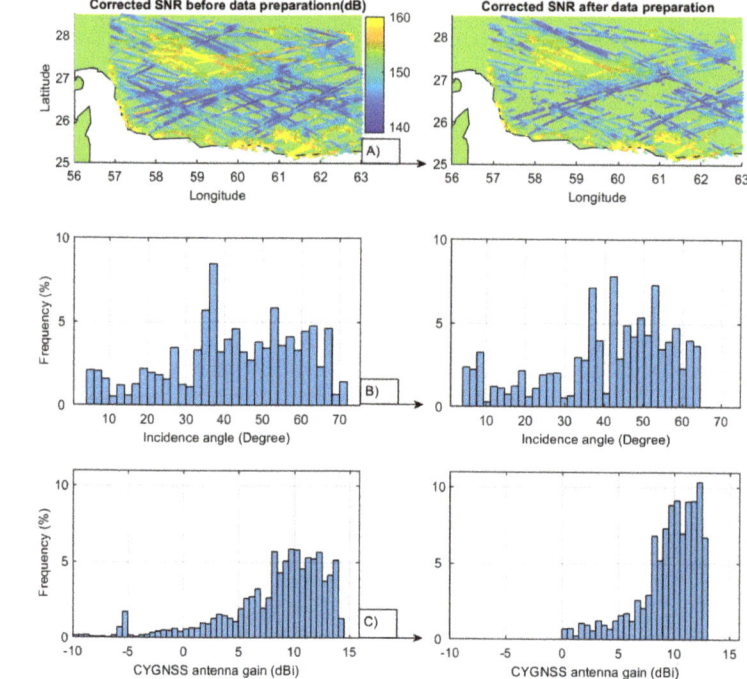

Figure 7. Statistical information of the data preparation step for the CYGNSS observations over a period of three days (13 January to 15 January 2020). (**A**) Corrected signal-to-noise ratio (SNR) before and after the preprocessing step. (**B**,**C**) The distribution of data according to incidence angle and antenna gain before and after data preparation.

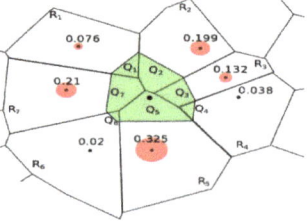

Figure 8. The natural neighbor interpolation method. The area of the colored circles are the interpolating weights. The shaded area is a new Voronoi element for the point to be interpolated [48].

For CYGNSS data interpolation over our region of interest, we generated a grid with the resolution of 0.1° along the geodetic longitude and latitude and applied the mentioned interpolation method. Figure 9 shows the data before and after gridding. As can be seen from the figure, the gridded data is more sensible compared to the satellite tracks representation. Since SNR was not equal in magnitude to the $SNR_{c\ dB}$, the observations and corrections made in Equation (4) resulted in magnitudes greater than 140 dB (Figure 7A). To see the anomalies of the corrected $SNR_{c\ dB}$ in a visually reasonable range [11], 140 dB was subtracted from the original $SNR_{c\ dB}$ values.

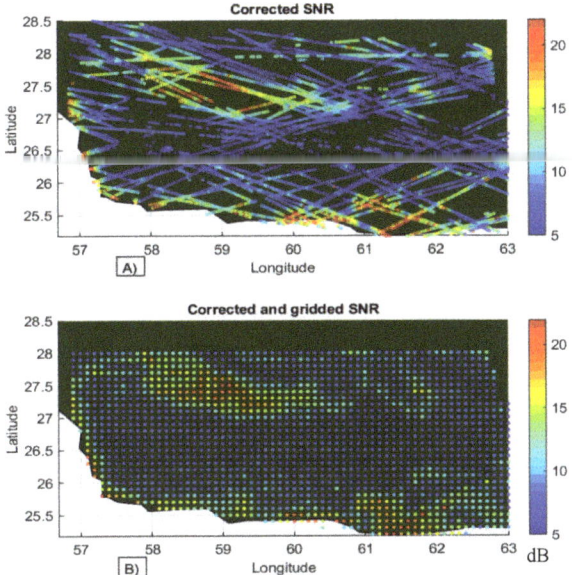

Figure 9. The outcome of the interpolation process for the corrected SNR ($SNR_{c\ dB}$) over the period of three days from 13 January to 15 January 2020. (**A**) Representation of the CYGNSS measurements along the satellite tracks, (**B**) the interpolated data at 0.1° × 0.1° grid points using the natural neighbor interpolation method.

4.4. Evaluation and Mapping

As can be seen in Figure 4B, the flood happened in the south and middle part of the Sistan and Baluchestan province in Iran. Figure 10 shows the flooded regions which were detected by CYGNSS observations overlaid on the MODIS image (Figure 4B) for verification. The figure contains three regions with significant SNR anomalies. The regions are labeled A, B, and C.

Region A in Figure 10 belongs to the Hamun-Jaz Murian depression in the southeast of Iran, placed between the Kerman province and Sistan and Baluchestan province. The shape of the depression or basin is oblong and enclosed by the mountains. There is a seasonal lake, Hamun, in the middle of the basin, which has been dry through the recent dry years. Although the Halil and Bampur rivers are the main sources of feeding for the basin, neither of both bring significant water to the basin to fill this lake, because the water is used for agricultural purposes on the way [49,50]. Moreover, the recent flood in January 2020 was unique in terms of flood volume over the last decade. The previous flood in this region happened in June 2007. Figure 11 demonstrates the capability of CYGNSS measurements in the detection and mapping of the flood over this depression.

To calculate the flooded areas using corrected SNR, a threshold was used to distinguish inundated from noninundated areas. A simple threshold method has been used in previous studies with monostatic and bistatic radars [11,18]. As is seen in Figures 10–13, observations with the values greater than 11 dB corresponded to the flooded areas. This threshold was used for the detection of inundation in this study. This value could be different in other regions. The roughness and vegetation could weaken the signals and change the threshold. The threshold used by [11] was 12dB for the medium-vegetation density and typical roughness.

As can be distinguished from flooded areas in Figures 11 and 12, the values of corrected SNR more than 11 dB have a high correlation with the satellite image in the inundated region. However, minor discrepancies could be related to georeferencing or interpolation errors. The overall evaluation of the results using the three days of CYGNSS data reports an acceptable performance for flood detection.

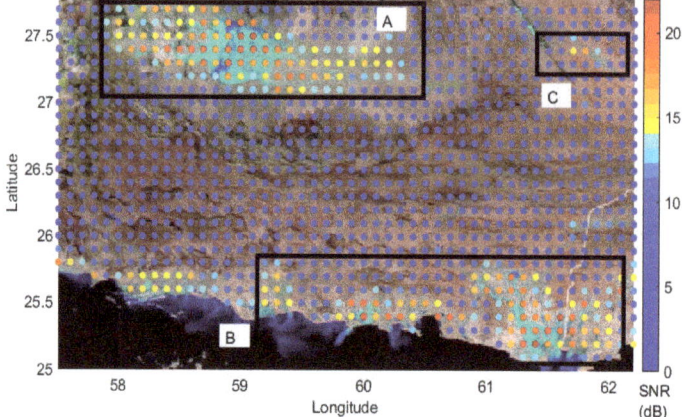

Figure 10. The georeferenced optical satellite imagery of the flood from MODIS (13 January 2020) overlaid by the corrected signal to noise ratio derived from CYGNSS observations (13 January to 15 January 2020). The regions labeled A, B, and C show significant SNR anomalies.

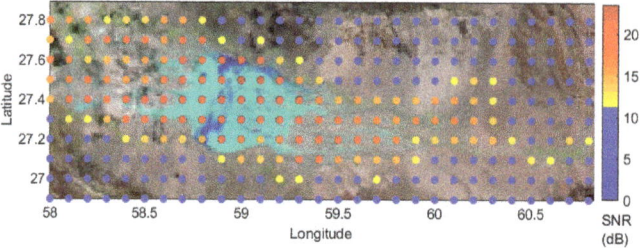

Figure 11. The georeferenced optical satellite imagery of the flood over the Hamun-Jaz Murian basin (region A in Figure 10) from MODIS (13 January 2020) overlaid by the corrected SNR derived from CYGNSS observations (13 January to 15 January 2020).

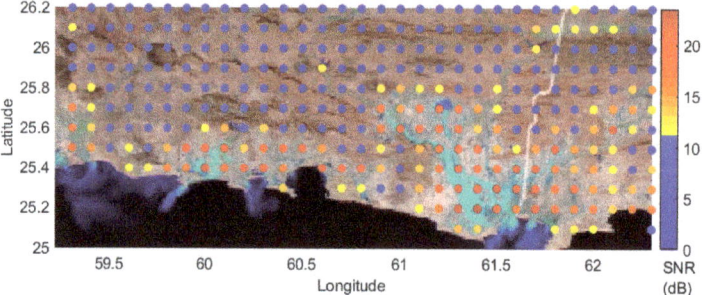

Figure 12. The georeferenced optical satellite imagery of the flood over coastlines and inundation of the nearby rivers (region B in Figure 10) from MODIS (13 January 2020) overlaid by the corrected SNR derived from CYGNSS observations (13 January to 15 January 2020).

Governments could use flood maps to establish the risk regions, safe evacuation options, and update the reaction plan. In the absence of promising and accurate flood maps, the development processes in or nearby the risk area are affected. The community lacks a tool to guide development to be more secure and to reduce future risks.

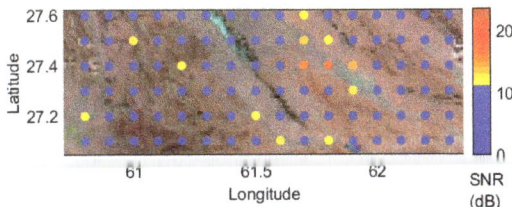

Figure 13. The georeferenced optical satellite imagery of the flood over the cities of Zaboli and Suran (region C in Figure 10) from MODIS (13 January 2020) overlaid by the corrected SNR derived from CYGNSS observations (13 January to 15 January 2020). The region includes a river and an inland water body.

We proceeded to map the detected inundation area. Google Maps was used here as an infrastructure which provides information about roads, cities, villages, etc. The derived data which shows flooded regions (data over 11 dB) was mapped on Google Maps. Figure 14 illustrates the three major regions of flood in Sistan and Baluchestan. Due to the flood in region A, corresponding to the Hamun-Jaz-Murian basin, the cities close to the basin, i.e., IranShahr, Eslam Abad, and Golmorti, and the roads between them, were affected. The area of this region is about 8706 square kilometers. Region B, which is close to the coastline and encompasses a few rivers, many cities, villages, farmlands, and roads, was also hit by the flood. The area of this region is about 9742 square kilometers. Region C, in close proximity to the region A, includes a river, an inland lake, and the cities Zaboli and Suran, which were affected. The area of this flooded region is about 1196 square kilometers. Therefore, based on the estimates from the CYGNSS observations, about 19644 square kilometers were affected by the flood in the south and middle parts of the Sistan and Baluchestan province. More severe impacts were seen in the regions close to the coastlines and nearby rivers.

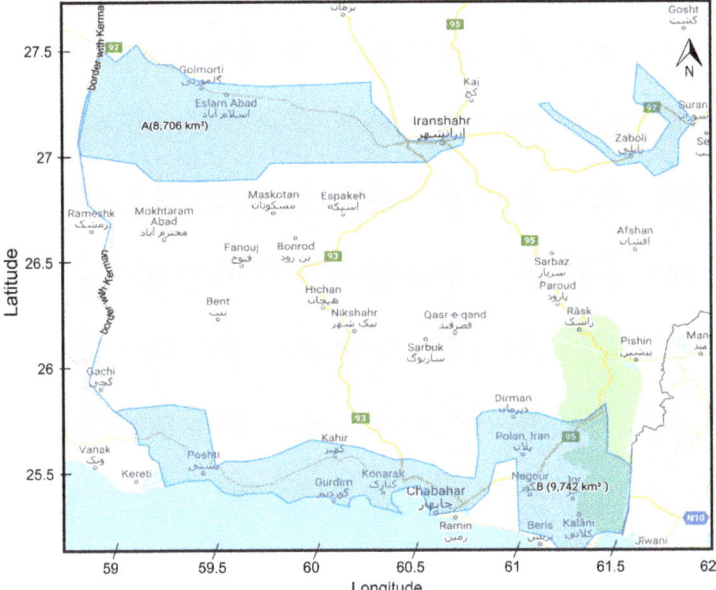

Figure 14. Map of the flooded regions laid over Google Maps. The blue-colored polygons show the boundary of the flooded areas, yellow lines are roads, and names of the cities are written on the map.

5. Summary and Conclusions

We applied the GNSS-R remote sensing technique based on a dataset of spaceborne observations of reflected GPS signals over the land to detect and map the recent flood in the southeastern part of Iran. The flood occurred in the Sistan and Baluchestan province after the heavy rain in mid-January 2020. The dataset used was acquired from the data products of the NASA CYGNSS mission. The main parameter of interest used in the analysis was the delay doppler map SNR, which was retrieved from the level-1 data product. First, a data preparation procedure was applied to remove outliers and discard low-quality data. In the next step, inverse bistatic radar formula was used to calculate the corrected SNR, which was closely related to surface reflectivity and hydrological conditions. The corrected SNR values were calibrated and interpolated to a regular grid over the study area. After calibration and gridding, the corrected SNR was verified with the MODIS optical image. A threshold of about 11 dB or more could be distinguished between the inundated and noninundated areas in the regions of interest. Finally, the flood-affected areas were mapped on Google Maps. The area of the flooded regions was estimated to be about 19,644 km^2 or 10.8% of the province. Many cities, roads, and other infrastructures were affected by the flood in these regions. The results indicate the regions close to depression, lakes, and coastal areas are at a high risk of flooding in this province. This study confirms that CYGNSS data is of value for hydrological investigations, particularly flood detection in the Sistan and Baluchestan province. Despite a relatively short revisit time of CYGNSS observations, the spatial resolution of the data products needs to be improved for mapping purposes. This issue could be addressed in future missions by, e.g., increasing the number of onboard processing channels, as well as by processing the reflected signals from other GNSS constellations such as GLONASS, Galileo, and BeiDou.

Author Contributions: Conceptualization, H.N., M.R.; Data curation, M.R., M.H.; Formal analysis, M.R.; Funding acquisition, M.R. and H.N.; Investigation, M.R. and M.H.; Methodology, M.R.; Software, M.R. and M.H.; Supervision, H.N.; Validation, M.R.; Visualization, M.R.; Writing—original draft, M.R.; Writing—review and editing, H.N. and M.H. All authors have read and agreed to the published version of the manuscript.

Funding: This research was funded by the Iran Ministry of Science, Research and Technology and the APC was funded by Norwegian University of Science and Technology.

Acknowledgments: The authors would like to acknowledge the valuable comments of the editors and two anonymous reviewers, which significantly improved the presentation and quality of this paper. Also, the authors would like to thank the NASA CYGNSS and MODIS science—operation teams for providing the data products which made this study possible.

Conflicts of Interest: The authors declare no conflict of interest.

References

1. Van Westen, C. Remote sensing for natural disaster management. *Int. Arch. Photogramm. Remote Sens.* **2000**, *33*, 1609–1617.
2. Salami, R.O.; von Meding, J.K.; Giggins, H. Vulnerability of Human Settlements to Flood Risk in the Core Area of Ibadan Metropolis, Nigeria. *Jàmbá J. Disaster Risk Stud.* **2017**, *9*, 1–14. [CrossRef] [PubMed]
3. ADPC; UNDP. *Integrated Flood Risk Management in Asia*; A Primer ADPC: Bangkok, Thailand, 2005.
4. Abidin, H.Z.; Andreas, H.; Gumiar, I.; Sidiq, T.P.; Gamal, M. Environmental impacts of land subsidence in urban areas of Indonesia. In *FIG Working Week*; TS 3—Positioning and Measurement: Sofia, Bulgaria, 2015.
5. Nash, L.L. *The Colorado River Basin and Climatic Change: The Sensitivity of Streamflow and Water Supply to Variations in Temperature and Precipitation*; US Environmental Protection Agency, Office of Policy, Planning, and Evaluation: Washington, DC, USA, 1993.
6. Houston, J. Variability of precipitation in the Atacama Desert: Its causes and hydrological impact. *Int. J. Clim. J. R. Meteorol. Soc.* **2006**, *26*, 2181–2198. [CrossRef]
7. Brivio, P.A.; Colombo, R.; Maggi, M.; Tomasoni, R. Integration of remote sensing data and GIS for accurate mapping of flooded areas. *Int. J. Remote Sens.* **2002**, *23*, 429–441. [CrossRef]
8. CNN. Available online: https://edition.cnn.com/2019/04/07/middleeast/iran-flood-fatalities/index.html. (accessed on 7 April 2019).

9. Oberstadler, R.; Hönsch, H.; Huth, D. Assessment of the mapping capabilities of ERS-1 SAR data for flood mapping: A case study in Germany. *Hydrol. Process.* **1997**, *11*, 1415–1425. [CrossRef]
10. Kuenzer, C.; Guo, H.; Huth, J.; Leinenkugel, P.; Li, X.; Dech, S. Flood mapping and flood dynamics of the Mekong Delta: ENVISAT-ASAR-WSM based time series analyses. *Remote Sens.* **2013**, *5*, 687–715. [CrossRef]
11. Chew, C.; Reager, J.T.; Small, E. CYGNSS data map flood inundation during the 2017 Atlantic hurricane season. *Sci. Rep.* **2018**, *8*, 1–8. [CrossRef]
12. Rajabi, M.; Amiri-Simkooei, A.R.; Asgari, J.; Nafisi, V.; Kiaei, S. Analysis of TEC time series obtained from global ionospheric maps. *J. Geomat. Sci. Technol.* **2015**, *4*, 213–224.
13. Rajabi, M.; Amiri-Simkooei, A.R.; Nahavandchi, H.; Nafisi, V. Modeling and Prediction of Regular Ionospheric Variations and Deterministic Anomalies. *Remote Sens.* **2020**, *12*, 936. [CrossRef]
14. Tabibi, S. Snow Depth and Soil Moisture Retrieval Using SNR-Based GPS and GLONASS Multipath Reflectometry. Ph.D. Thesis, University of Luxembourg, Luxembourg, 2016.
15. Li, W.; Cardellach, E.; Fabar, F.; Rius, A.; Ribó, S.; Martín-Neira, M. First spaceborne phase altimetry over sea ice using TechDemoSat-1 GNSS-R signals. *Geophys. Res. Lett.* **2017**, *44*, 8369–8376. [CrossRef]
16. Chew, C.; Small, E. Soil moisture sensing using spaceborne GNSS reflections: Comparison of CYGNSS reflectivity to SMAP soil moisture. *Geophys. Res. Lett.* **2018**, *45*, 4049–4057. [CrossRef]
17. Wu, X.; Li, Y.; Xu, J. Theoretical study on GNSS-R vegetation biomass. In Proceedings of the 2012 IEEE International Geoscience and Remote Sensing Symposium, Munich, Germany, 12 November 2012.
18. Wan, W.; Liu, B.; Zeng, Z.; Chen, X. Using CYGNSS data to monitor China's flood inundation during typhoon and extreme precipitation events in 2017. *Remote Sens.* **2019**, *11*, 854. [CrossRef]
19. Hoseini, M.; Asgarimehr, M.; Zavorotny, V.; Nahavandchi, H.; Ruf, C.; Wickert, J. First evidence of mesoscale ocean eddies signature in GNSS reflectometry measurements. *Remote Sens.* **2020**, *12*, 542. [CrossRef]
20. Clarizia, M.P.; Ruf, C.S. Wind speed retrieval algorithm for the Cyclone Global Navigation Satellite System (CYGNSS) mission. *IEEE Trans. Geosci. Remote Sens.* **2016**, *54*, 4419–4432. [CrossRef]
21. Liu, B.; Wan, W.; Hong, Y. Can the Accuracy of Sea Surface Salinity Measurement Be Improved by Incorporating Spaceborne GNSS-Reflectometry? *IEEE Geosci. Remote Sens. Lett.* **2020**. [CrossRef]
22. Zavorotny, V.U.; Gleason, S.; Cardellach, E.; Camps, A. Tutorial on remote sensing using GNSS bistatic radar of opportunity. *IEEE Geosci. Remote Sens. Mag.* **2014**, *2*, 8–45. [CrossRef]
23. Eroglu, O.; Kurum, M.; Boyd, D.; Gurbuz, A.C. High Spatio-Temporal Resolution CYGNSS Soil Moisture Estimates Using Artificial Neural Networks. *Remote Sens.* **2019**, *11*, 2272. [CrossRef]
24. Teunissen, P.; Montenbruck, O. *Springer Handbook of Global Navigation Satellite Systems*; Springer: Cham, Switzerland, 2017.
25. Guerriero, L.; Pierdicca, N.; Egido, A.; Caparrini, M.; Paloscia, S.; Santi, E.; Floury, N. Modeling of the GNSS-R signal as a function of soil moisture and vegetation biomass. In Proceedings of the 2013 IEEE International Geoscience and Remote Sensing Symposium-IGARSS, Melbourne, VIC, Australia, 21–26 July 2013.
26. Masters, D.; Axelrad, P.; Katzberg, S. Initial results of land-reflected GPS bistatic radar measurements in SMEX02. *Remote Sens. Environ.* **2004**, *92*, 507–520. [CrossRef]
27. Ruf, C.; Chang, P.S.; Clarizia, M.P.; Gleason, S.; Jelenak, Z. *CYGNSS Handbook*; Michigan Pub: Ann Arbor, MI, USA, 2016; p. 154.
28. Ruf, C.S.; Atlas, R.; Chang, P.S.; Clarizia, M.P.; Garrison, J.; Gleason, S.; Katzberg, S.; Jelenak, Z.; Johnson, J.; Sharanya, J.; et al. New ocean winds satellite mission to probe hurricanes and tropical convection. *Bull. Am. Meteorol. Soc.* **2016**, *97*, 385–395. [CrossRef]
29. Ruf, C.S.; Chew, C.; Lang, T.; Morris, M.; Nave, K.; Ridley, A.; Balasubramaniam, R. A new paradigm in earth environmental monitoring with the CYGNSS small satellite constellation. *Sci. Rep.* **2018**, *8*, 1–13. [CrossRef]
30. Zribi, M.; Gorrab, A.; Baghdadi, N.; Lili-Chabaane, Z.; Mougenot, B. Influence of radar frequency on the relationship between bare surface soil moisture vertical profile and radar backscatter. *IEEE Geosci. Remote Sens. Lett.* **2013**, *11*, 848–852. [CrossRef]
31. Camps, A.; Park, H.; Pablos, M.; Foti, G.; Gommenginger, C.; Liu, P.W.; Jugge, J. Sensitivity of GNSS-R spaceborne observations to soil moisture and vegetation. *IEEE J. Sel. Top. Appl. Earth Obs. Remote Sens.* **2016**, *9*, 4730–4742. [CrossRef]
32. Chew, C.; Shah, R.; Zuffada, Z.; Hajj, G.; Masters, D.; Mannuci, A. Demonstrating soil moisture remote sensing with observations from the UK TechDemoSat-1 satellite mission. *Geophys. Res. Lett.* **2016**, *43*, 3317–3324. [CrossRef]

33. Gerlein-Safdi, C.; Ruf, C.S. A CYGNSS-based algorithm for the detection of inland waterbodies. *Geophys. Res. Lett.* **2019**, *46*, 12065–12072. [CrossRef]
34. Morris, M.; Chew, C.; Reager, J.; Shah, R.; Zuffada, C. A novel approach to monitoring wetland dynamics using CYGNSS: Everglades case study. *Remote Sens. Environ.* **2019**, *233*, 111417. [CrossRef]
35. Amineh, Z.B.A.; Hashemian, S.J.A.-D.; Magholi, A. Integrating Spatial Multi Criteria Decision Making (SMCDM) with Geographic Information Systems (GIS) for delineation of the most suitable areas for aquifer storage and recovery (ASR). *J. Hydrol.* **2017**, *551*, 577–595. [CrossRef]
36. World Data. Available online: https://www.worlddata.info (accessed on 1 February 2020).
37. Huffman, G.J.; Bolvin, D.T.; Nelkin, E.J. Integrated Multi-satellitE Retrievals for GPM (IMERG) technical documentation. *NASA/GSFC Code* **2015**, *612*, 47.
38. Lindsey, R.; Herring, D.; Abbott, M.; Conboy, B.; Esaias, W. *MODIS Brochure*; Goddard Space Flight Center: Greenbelt, MD, USA, 2013.
39. Flash Flooding in Iran. 2020. Available online: https://earthobservatory.nasa.gov/images/146150/flash-flooding-in-iran (accessed on 1 February 2020).
40. Eaves, J.L. Introduction to radar. In *Principles of Modern Radar*; Van Nostrand Reinhold; Springer: New York, NY, USA, 1987; pp. 1–27.
41. Bruder, J.A. IEEE Radar standards and the radar systems panel. *IEEE Aerosp. Electron. Syst. Mag.* **2013**, *28*, 19–22. [CrossRef]
42. Nghiem, S.V.; Zuffada, C.; Shah, R.; Chew, C.; Lowe, S.T.; Mannucci, A.J.; Cardellach, E.; Brakenridge, G.R.; Geller, G.; Rosenqvist, A. Wetland monitoring with global navigation satellite system reflectometry. *Earth Space Sci.* **2017**, *4*, 16–39. [CrossRef]
43. Masters, D. Surface Remote Sensing Applications of GNSS Bistatic Radar: Soil Moisture and Aircraft Altimetry. Ph.D. Thesis, University of Colorado, Boulder, CO, USA, 2004.
44. Wang, T.; Ruf, C.S.; Block, B.; McKague, D.; Gleason., S. Characterization of GPS L1 EIRP: Transmit power and antenna gain pattern. In Proceedings of the 31st ION GNSS, Miami, FL, USA, 24–28 September 2018.
45. Basis, A.T. Cyclone Global Navigation Satellite System (CYGNSS). Prepared by: Maria Paola Clarizia, University of Michigan, Valery Zavorotny, NOAA. 2015.
46. Sibson, R. *A Brief Description of Natural Neighbor Interpolation (Chapter 2)*; Barnett, V., Ed.; Interpolating Multivariate Data; John Wiley: Chichester, UK, 1981; pp. 21–36.
47. Voronoi, G. Nouvelles applications des paramètres continus à la théorie des formes quadratiques. Deuxième mémoire. Recherches sur les parallélloèdres primitifs. *J. für die reine und angewandte Mathematik (Crelles J.)* **1908**, *1908*, 198–287. [CrossRef]
48. Tsidaev, A. Parallel algorithm for natural neighbor interpolation. In Proceedings of the 2nd Ural Workshop on Parallel, Distributed, and Cloud Computing for Young Scientists, Yekaterinburg, Russia, 6 October 2016.
49. Harrison, J. The Jaz Murian depression, Persian Baluchistan. *Geogr. J.* **1943**, *101*, 206–225. [CrossRef]
50. Frs, N.F. From Musandam to the Iranian Makran. *Geogr. J.* **1975**, *141*, 55–58. [CrossRef]

© 2020 by the authors. Licensee MDPI, Basel, Switzerland. This article is an open access article distributed under the terms and conditions of the Creative Commons Attribution (CC BY) license (http://creativecommons.org/licenses/by/4.0/).

Article

Flood Monitoring in Vegetated Areas Using Multitemporal Sentinel-1 Data: Impact of Time Series Features

Viktoriya Tsyganskaya [1,*], Sandro Martinis [2] and Philip Marzahn [1]

1   Department of Geography, Ludwig Maximilian University of Munich, Luisenstr. 37, Munich 80333, Germany
2   German Aerospace Center (DLR), German Remote Sensing Data Center (DFD), Oberpfaffenhofen, 82234 Wessling, Germany
*   Correspondence: tsyganskaya.viktoriya@gmx.de

Received: 10 July 2019; Accepted: 12 September 2019; Published: 18 September 2019

Abstract: Synthetic Aperture Radar (SAR) is particularly suitable for large-scale mapping of inundations, as this tool allows data acquisition regardless of illumination and weather conditions. Precise information about the flood extent is an essential foundation for local relief workers, decision-makers from crisis management authorities or insurance companies. In order to capture the full extent of the flood, open water and especially temporary flooded vegetation (TFV) areas have to be considered. The Sentinel-1 (S-1) satellite constellation enables the continuous monitoring of the earths surface with a short revisit time. In particular, the ability of S-1 data to penetrate the vegetation provides information about water areas underneath the vegetation. Different TFV types, such as high grassland/reed and forested areas, from independent study areas were analyzed to show both the potential and limitations of a developed SAR time series classification approach using S-1 data. In particular, the time series feature that would be most suitable for the extraction of the TFV for all study areas was investigated in order to demonstrate the potential of the time series approaches for transferability and thus for operational use. It is shown that the result is strongly influenced by the TFV type and by other environmental conditions. A quantitative evaluation of the generated inundation maps for the individual study areas is carried out by optical imagery. It shows that analyzed study areas have obtained Producer's/User's accuracy values for TFV between 28% and 90%/77% and 97% for pixel-based classification and between 6% and 91%/74% and 92% for object-based classification depending on the time series feature used. The analysis of the transferability for the time series approach showed that the time series feature based on VV (vertical/vertical) polarization is particularly suitable for deriving TFV types for different study areas and based on pixel elements is recommended for operational use.

Keywords: flood mapping; temporary flooded vegetation (TFV); Sentinel-1; time series data; Synthetic Aperture Radar (SAR)

---

1. Introduction

Flood events are the most frequent and widespread natural hazards worldwide and can have devastating economic, social, and environmental impacts [1,2]. Precise and timely information on the extent of flooding is therefore essential for various institutions such as relief organizations, decision-makers of crisis management authorities or insurance companies [3].

Satellite Synthetic Aperture Radar (SAR) is particularly suitable for flood mapping, as this tool supports the large-scale, cross-border detection of the affected area independent of illumination and weather conditions [4–6]. The decisive advantage, however, is that in addition to open water surfaces, temporary flooded vegetation (TFV) can also be detected in dependency of system and environmental parameters [7]. TFV are areas where water bodies temporarily occur underneath the vegetation [8]. To avoid underestimations of the flooding, the derivation of both classes is essential to cover the entire flood extent.

Smooth open water is characterized by low SAR backscatter values due to its specular surface. In comparison, TFV shows a significant increase in backscatter, especially in the VV (vertical/vertical) polarization, which is caused by the complex double- or multi-bounce interaction between smooth open water surfaces and the structure of vegetation (e.g., tree trunks, stems) [8–10].

Most approaches are based on the backscatter intensity allowing the detection of TFV by the identification of increased backscatter values compared to other objects (e.g., [4,11–15]). Others utilize polarimetric decomposition and/or interferometric SAR (InSAR) coherence [7,16–19] to reduce confusion with urban areas or to minimize the misclassification of shadowed regions as non-flooded vegetation. Polarimetric decompositions, such as Freeman–Durden, Yamaguchi four-component, Cloude–Pottier, or m-chi decompositions, have all been demonstrated to be suitable for the extraction of TFV [7,20–28]. However, the availability of full polarimetric data is often limited regarding the extent and temporal coverage.

In the literature, various methods for deriving the flood extent based on SAR data can be found depending on the task, polarization modes, phase information, as well as spatial or temporal resolution of the satellite sensor [29]. Some of them include, for example, visual interpretation [30], histogram thresholding approaches [31,32], image texture-based methods [33], Markov Random Field modeling [34], or Wishart classifications [17,35,36], which are mostly applied on single images. Change detection techniques in combination with algorithms, such as manual or automatic thresholding [21,37] and fuzzy logic [38,39], allow the extraction of potential changes between two images acquired under dry and flood conditions. Change detection methods are often carried out by using absolute backscatter values [40], which do not consider the chance intensity of backscatter values within vegetation. This can lead to classification errors in regions with high vegetation growth variability or with different vegetation types.

A few advanced techniques, among other machine learning techniques [11], decision tree [41], or rule-based classification [13,42] use satellite time series [4,40,43–47] or multi-dates [11,38,48–52], which allow the inclusion of multi-temporal, -polarized or/and ancillary information for the extraction of temporary open water (TOW) and TFV classes. Thereby, seasonal or annual fluctuations of backscatter and multiple observations of the same area can be used to improve the reliability of mapping the flood extent or even the flood dynamics [43]. Moreover, the use of multitemporal approaches has been in the past limited due to the low availability of corresponding SAR data. However, since October 2014, the Sentinel-1 (S-1) satellite constellation has continuously and systematically captured the earths surface with C-band SAR data at short repetition time, enabling the use of SAR multi-temporal data for systematic and operational flood monitoring. Using this data source, Tsyganskaya et al., [8] recently showed a time series approach for the detection of TFV.

This study aims to show the potential of the SAR time series approach proposed in Tsyganskaya et al., [8] regarding the extraction of the entire flood extent with the focus on TFV for two independent study areas in Greece/Turkey and China. The main focus of the study is to demonstrate the impact of the time series features on the classification results and to show their potential for operational use. The objectives in detail are as follows:

- to investigate the relevance of polarization and time series features for the derivation of TFV with respect to vegetation types in both study areas;
- to examine if the relevant time series features for the analyzed study areas correspond to the relevant time series features of the previous study area (Namibia) in [8], despite the occurrence of different vegetation types and
- to identify a single time series feature that is relevant for the extraction of different TFV types and for all study areas in order to demonstrate the potential for the transferability and operational use of this time series approach.

2. Materials and Methods

2.1. Study Areas and Available Data Sets: Greece/Turkey and China

Besides the study area in Namibia described in [8], two further study areas with different vegetation types were used for the impact analysis of time series features for the extraction of TFV. Compared to [8] and to each other, both study areas have different vegetation types which are described in this section. One of the study areas is part of the Evros catchment, located at the border between Greece and Turkey (Figure 1a) is one of the study areas.

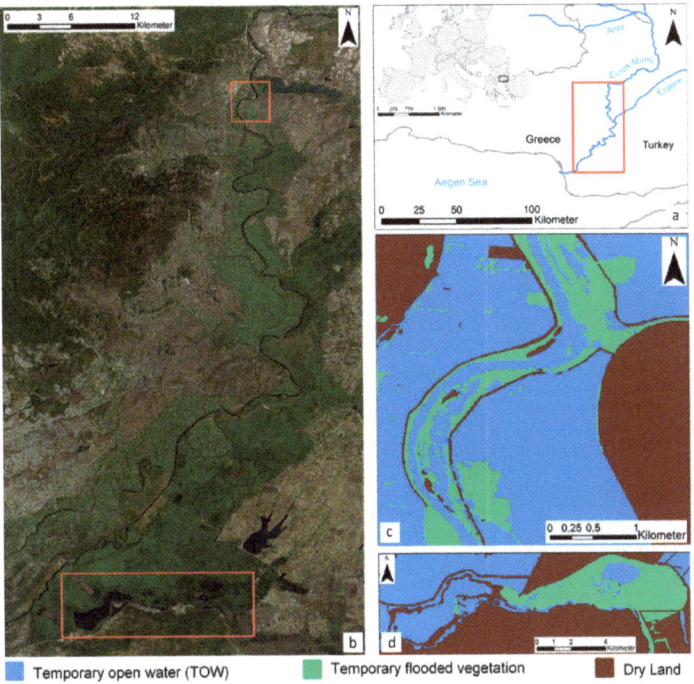

Figure 1. Overview map with the location of the study area (red rectangle) in Greece/Turkey (**a**). Study area in Greece/Turkey (Satellite data: Esri, DigitalGlobe, GeoEye, Earthstar Geographics, CNES/Airbus DS, USDA, USGS, AeroGRID, IGN, and the GIS Use Community) (**b**). The red rectangles represent the reference mask extents for northern (**c**), and southern (**d**) areas. The reference masks based on World-View 2 data (March 11, 2015) (**c**) and RapidEye (April 4, 2015) data (**d**).

With a length of 515 km and a basin of about 52,900 km^2, the Evros river represents the second largest river in Eastern Europe. The period of highest discharge usually occurs between December and April. Several severe, large-scale floods frequently hit the catchment area, particularly in the southern part. The focus of the study lies on a flood event in spring 2015. The flood-affected areas consisted of farmland and forested areas where the water remained for over several weeks. For the analysis and classification, a time series of 60 dual-polarized S-1 scenes was used, which were acquired under the same orbital conditions between October 2014 and December 2016 (Table 1). Five S-1 images covered the flood event. Figure 1b shows the extent of the northern and southern areas of interest, where two scenes, acquired on March 12, 2015 and April 5, 2015 were considered for classification as flood event images due to their temporal proximity to reference data for the two different areas. The northern area is dominated by TFV consisting of deciduous forest, whereby the majority of agricultural fields are entirely inundated during the flood event. In the southern part of the study area, TFV occurs mostly in high grassland areas.

Table 1. Acquisition dates of the Sentinel (S-1) satellite data used for Greece/Turkey. The scenes acquired on March 12, 2015 and April 5, 2015 (highlighted with blue background) were used as flood event images for two different parts (northern and southern) of the Greece/Turkey study area.

No	Date	No	Date	No	Date	No	Date
1	October 19, 2014	16	June 16, 2015	31	January 6, 2016	46	July 16, 2016
2	October 31, 2014	17	June 28, 2015	32	January 18, 2016	47	July 28, 2016
3	November 24, 2014	18	July 10, 2015	33	January 30, 2016	48	August 9, 2016
4	December 6, 2014	19	July 22, 2015	34	February 11, 2016	49	August 21, 2016
5	December 18, 2014	20	August 15, 2015	35	February 23, 2016	50	September 2, 2016
6	December 30, 2014	21	August 27, 2015	36	March 6, 2016	51	September 14, 2016
7	January 11, 2015	22	September 8, 2015	37	March 18, 2016	52	September 26, 2016
8	February 4, 2015	23	September 20, 2015	38	March 30, 2016	53	October 2, 2016
9	February 16, 2015	24	October 2, 2015	39	April 11, 2016	54	October 14, 2016
10	March 12, 2015	25	October 14, 2015	40	April 23, 2016	55	October 26, 2016
11	March 24, 2015	26	October 26, 2015	41	May 5, 2016	56	November 7, 2016
12	April 5, 2015	27	November 19, 2015	42	May 17, 2016	57	November 19, 2016
13	April 17, 2015	28	December 1, 2015	43	May 29, 2016	58	December 1, 2016
14	May 11, 2015	29	December 13, 2015	44	June 10, 2016	59	December 13, 2016
15	June 4, 2015	30	December 25, 2015	45	July 4, 2016	60	December 25, 2016

Validation of the classification was performed based on two reference masks (Figure 1c,d). The generation of the reference data was carried out by visual interpretation and manual digitalization of high-resolution optical WorldView-2 (Figure 2) and RapidEye (Figure 3) images, acquired on March 11, 2015 and April 4, 2015, respectively. Although the radar data and the optical image have a temporal shift of one and two days respectively, no changes in the flood extent could be observed.

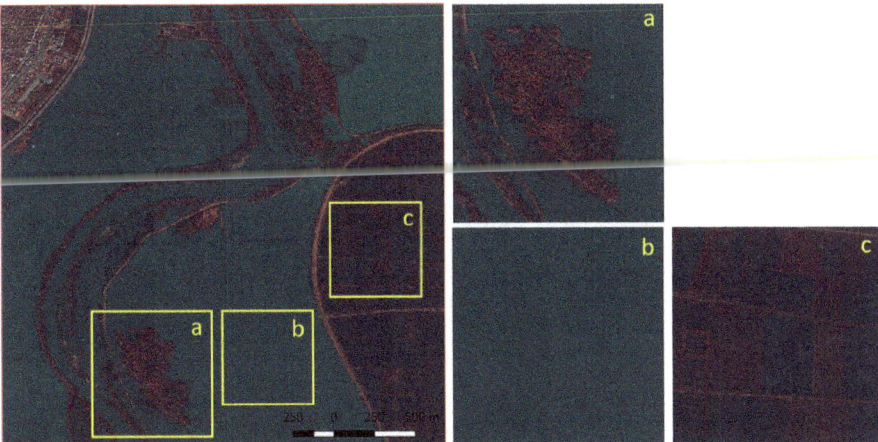

Figure 2. High-resolution false-color (NIR (near infrared), green, blue) WorldView2 image (March 11, 2015) (© European Space Imaging/DigitalGlobe) for northern Greece/Turkey containing temporary flooded vegetation (TFV) (**a**), temporary open water (TOW) (**b**), and dry land (DL) (**c**).

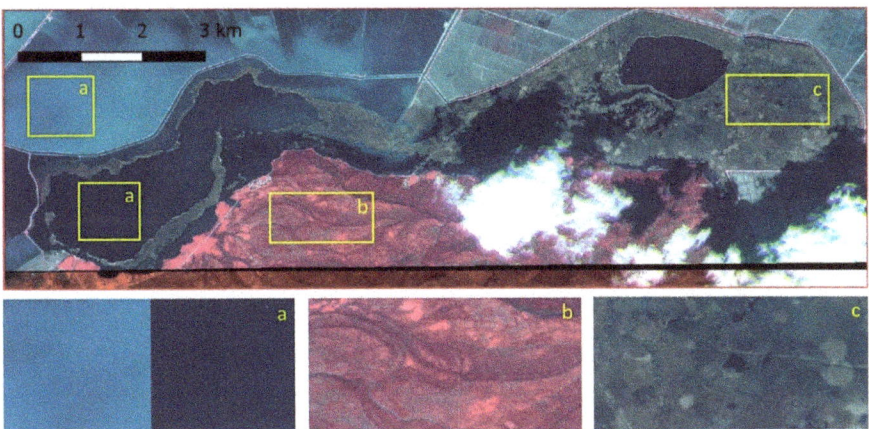

Figure 3. High-resolution false-color (NIR, green, blue) RapidEye image (April 04, 2015). (© Planet Labs Inc.) for southern Greece/Turkey containing TOW (**a**), DL (**b**), and TFV (**c**).

The second test case is the Dong Ting Lake, which is the second largest lake in China, located in the Hunan Province (Figure 4a). It is a flood-basin of the Yangtze River and thus varies seasonally in size. During the annual floods, it can expand to 2691 km^2 three times its size compared to the dry season. A flood event in summer 2017 (Figure 4b) was chosen. For this study, 38 dual-polarized S-1 scenes were used, which were acquired between October 2016 and February 2018 (Table 2). The scene acquired on June 28, 2017 is characterized by the largest flood extent. In addition, the selection of the analyzed flood image is carried out due to the temporal proximity to the reference data. Comparable to the study area in Greece/Turkey, the generation of the reference flood mask was carried out using a high-resolution optical Sentinel-2 (S-2) image (Figure 5), which was acquired on June 27, 2017 (Table 2). For the derivation of the reference mask, all bands with a resolution of 10 m and their combinations of the S-2 scene were used. The reference extent and the digitalized reference mask is shown in Figure 4c. No changes in flood extent were observed between the analyzed flood image and the optical scene.

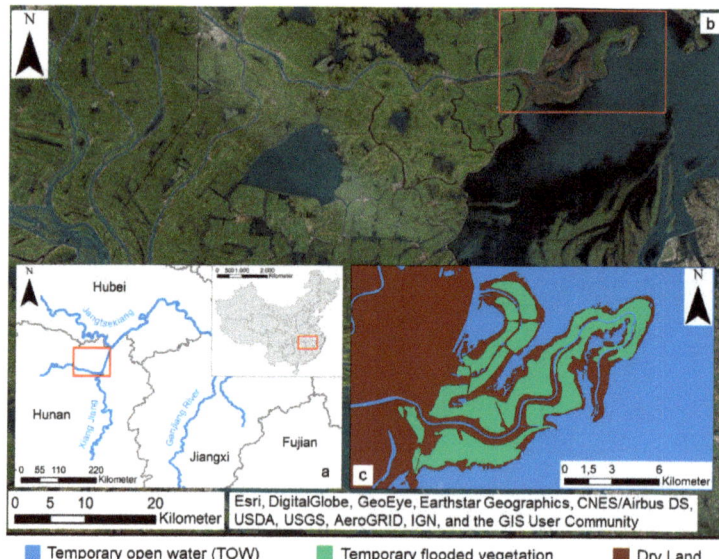

Figure 4. Overview map with the location of the study area (red rectangle) in China (**a**), study area in China (**b**). The red rectangle represents the reference mask extent (**c**), which was derived from S-2 data (June 27, 2017).

Table 2. Acquisitions dates of the S-1 satellite data used for China. The scene acquired on June 28, 2017 (highlighted with blue background) was used within this study as a flood event image.

No	Date	No	Date	No	Date	No	Date
1	October 19, 2016	11	February 16, 2017	21	July 10, 2017	31	November 19, 2017
2	October 31, 2016	12	February 28, 2017	22	July 22, 2017	32	December 1, 2017
3	November 12, 2016	13	March 12, 2017	23	August 3, 2017	33	December 13, 2017
4	November 24, 2016	14	March 24, 2017	24	August 15, 2017	34	December 25, 2017
5	December 6, 2016	15	April 5, 2017	25	August 27, 2017	35	January 6, 2018
6	December 18, 2016	16	April 17, 2017	26	September 8, 2017	36	January 30, 2018
7	December 30, 2016	17	April 29, 2017	27	October 2, 2017	37	February 11, 2018
8	January 11, 2017	18	June 11, 2017	28	October 14, 2017	38	February 23, 2018
9	January 23, 2017	19	June 4, 2017	29	October 26, 2017		
10	February 4, 2017	20	June 28, 2017	30	November 7, 2017		

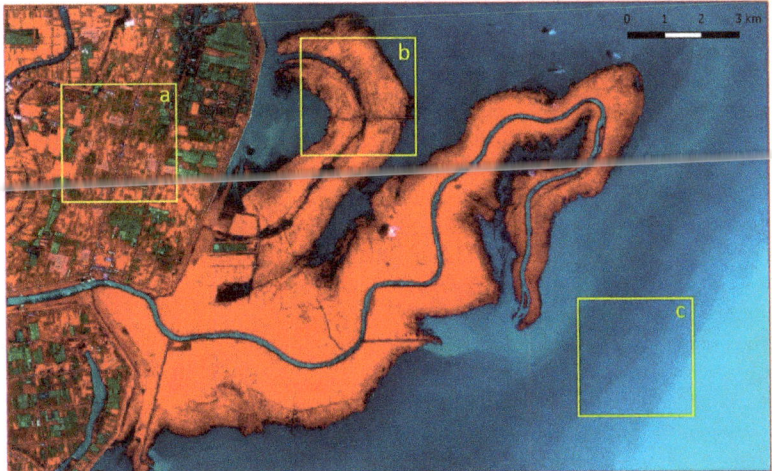

Figure 5. High-resolution false-color (NIR, green, blue) Sentinel-2 image (June 27, 2017) for southern Greece/Turkey containing DL (**a**), TFV (**b**) and TOW (**c**).

According to the ESA (European Space Agency) acquisition plan for S-1 constellation, dual-polarized images of the S-1A with an interval of 12 days were available for the study areas and the analyzed acquisition period. These data sets are processed and provided by ESA as Ground Range Detected High Resolution (GRDH) products. According to [8], an automated preprocessing of the time series data for both areas and both polarizations VV and VH (vertical/horizontal) was carried out in several steps. The characteristics of all used S-1 data are listed in Table 3.

Table 3. Characteristics of the S-1 data used.

Sensor Properties	Values
Wavelength Mode	Interferometric Wide Swath (IW)
Polarization	VV – VH
Frequency	C-band (GHz)
Resolution	20 × 22 m (az. × gr. range)
Pixel spacing	10 × 10 m (az. × gr. range)
Inc. angle	30.5°–46.3°
Orbit	Ascending
Product-level	Level-1 (Ground Range Detected High Resolution (GRDH))

In addition to the satellite data, ancillary information was used to avoid misclassification of the desired flood-related classes [8] in both study areas. Initially, permanent open water (POW) surfaces were identified by the 30 m SRTM Water Body Data (SWBD) [53]. Besides POW, information regarding urban areas and topography was integrated by using the Global Urban Footprint (GUF) and Height Above Nearest Drainage (HAND) index. Urban areas, as well as TFV, are characterized by strong double- and multiple-bounce backscattering effects. Therefore, the GUF [54] mask is used to identify and exclude urban areas. In order to prevent misclassifications in elevated areas, the HAND index [55] was used to identify and exclude areas with an elevation of greater than 20 m above the nearest water network [56].

2.2. Methodology

The foundation for the analysis of the impact of time series features on the flood extent derivation is the SAR time series approach [8]. This method provides the opportunity to generate flood maps including the classes TOW, TFV and Dry Land (DL) based on different time series features. The main steps of the classification process for the generation of the flood maps are summarized below.

The S-1 time series data presented in Section 2.1 provide a foundation for the generation of two independent multi-temporal layer stacks for each polarization and for each study area via an image preprocessing step. In addition, three further multitemporal stacks are generated, including the combinations of polarization (VV + VH, VV − VH, VV/VH). The ratio between VV and VH was calculated by using linear units, whereby for the elimination of outliers (e.g., due to speckle) only the values within the 5th and 95th percentiles were considered.

Based on findings about characteristics and patterns for TOW and TFV [8], the time series features were determined using the z-transform of the backscatter values for each pixel over the time series and for the available multi-temporal layer stack. The z-transform allows the normalization of the backscatter values over the time series and ensures that the image elements of the analyzed flood scene are comparable with each other when considering the individual seasonal fluctuation of vegetation. The normalized time series features are termed as Z-Score. In relation to the polarizations, the time series features are Z-Score VV, Z-Score VH, the combination of polarization, Z-Score VV + VH, Z-Score VV − VH and the ratio Z-Score VV/VH, which represent the foundation for the derivation of TOW and TFV [8]. Based on training data, two time series features with the highest contribution for the derivation of the desired classes (TOW and TFV) were identified for each study area by using Random Forest Algorithm. These features were used in the last step of the classification approach.

The classification of the flood-related classes was carried out based on pixels and objects. The K-means clustering algorithm [57] was applied for the generation of clusters using SAR time series data and the analyzed flood image. The resulting multitemporal and spatial cluster images were intersected with each other to combine multi-temporal and spatial information. The time series features, which are based on pixels, were merged with the segmented image by averaging the corresponding values within each object.

The hierarchical thresholding approach is the last step in the classification process chain, which allows a successive derivation of the desired classes based on the image elements (pixels or segments). First of all, the permanent water surfaces are identified by the SWBD mask (see Section 2.1). On the remaining image elements, the TOW, TFV and Dry Land (DL) are then derived consecutively using the relevant time series features. The corresponding thresholds were determined automatically using the decision tree classifier and the above-mentioned training data. For further details and an in-depth explanation see [8]. The implementation of the entire approach was done in Python. The quantification of the classification accuracy of TOW, TFV, and DL was carried out using overall accuracy (OA), producer accuracy (PA), user accuracy (UA), and kappa index (K).

In order to investigate the impact of time series features for the derivation of TFV and to show the potential of time series features for the transferability to different study areas and thus for the operational use, a statistical comparison of the classification accuracies based on different study areas and TFV types was performed. Thereby, a single time series feature was searched for, which allows the extraction of different types of TFV with high accuracy for all study areas. Besides the study areas described in Section 2.1, the study area in Namibia, which was analyzed in Tsyganskaya et al., [8], was also used for the investigation.

In order to identify a single time series feature, which can then be used for all study areas, the mean value and the coefficient of variation (CV) were determined based on user accuracy (UA) and producer accuracy (PA) for each time series feature. The mean value represents the total accuracy for the TFV classification of all study areas and for each time series feature. The higher the mean value, the more relevant is the time series feature for all study areas. In addition, the CV was used to quantify the variance of the data relative to the mean. The variance of the values is particularly relevant with regard to the outliers in UA and PA. The lower the CV, the lower the distance/scatter between the classification accuracies of the study areas for a time series feature and the more relevant is a time series feature for all study areas and thus for their transferability. Thus, the two statistical quantities allow the comparison between the classification results and the time series features for all study areas simultaneously.

3. Results and Discussion

3.1. Level of Contribution of the Time Series Features

Based on the knowledge about the characteristics and patterns of the time series data for the two polarizations and their combinations, time series features were derived [8]. Using training data for each of the study areas, those features were examined for their relevance to the extraction of the desired classes, TOW and TFV. For this purpose, similarly large training data sets were derived from optical images for the three classes TOW, TFV, and DL and for each of the study areas. Using Random Forest Algorithm, it was identified which time series features provide the highest contribution to the classification results and are, therefore, most relevant for the derivation of flood-related classes. Furthermore, it was examined whether a single time series feature or a combination of time series features enables the highest classification accuracy for the two desired classes. As a result, redundant information can be sorted out and the information required for classification identified.

Tables 4 and 5 show the level of contribution of the analyzed time series features for the derivation of the above-mentioned classes and the three study areas. Z-Score VV + VH represents the time series feature with the highest contribution for TOW for all study areas (Table 5). This can be explained by the fact that for both VV and VH polarizations the backscatter values decrease at the analyzed date of the flood [8]. An example of the backscatter value decrease in VV and VH is shown in Figure 6, which was created for a TOW segment from the validation data. Z-Score VV + VH combines both sources of information, whereby the decrease of the backscatter values is intensified. Figure 2b shows the enlarged view of the high-resolution WorldView2 image, demonstrating the presence of the corresponding TOW at the analyzed flood date. Further examples of TOW and the analyzed extent of study areas in southern Greece/Turkey and China are demonstrated in Figures 3a and 5c. In addition to both polarizations, the multi-temporal behavior of the Normalized Difference Vegetation Index (NDVI) values for the same time-period is displayed in Figure 6 serving as a comparison to the SAR time series data. NDVI values were derived from the LANSAT 8 data sets. Despite the cloud-related data gap for 2014, the beginning of 2015 and the beginning of 2016, a strong decrease in the NDVI values for the analyzed date of the flood event can be observed. In combination with the decrease of the backscatter values, this confirms the occurrence of water at the analyzed date. For the classification of TFV, different time series features with the highest contribution or relevance can be identified for the individual study areas. The differences can be attributed to the different types of vegetation and the different environmental conditions in the study areas.

Table 4. Random forest importance analysis of the time series features for TOW and individual study areas. The rows highlighted in blue represent the time series features with the highest contribution (importance).

	Southern Greece/Turkey (%)	Northern Greece/Turkey (%)	China (%)
Z-Score VV	35.64	29.26	31.35
Z-Score VH	15.51	23.07	24.61
Z-Score VV + VH	39.44	35.3	33.37
Z-Score VV − VH	5.91	7.33	3.87
Z-Score VV/VH	3.51	5.08	6.81

Table 5. Random forest importance analysis of the time series features for TFV and individual study areas. The rows highlighted in green represent the time series features with the highest contribution (importance).

	Southern Greece/Turkey (%)	Northern Greece/Turkey (%)	China (%)
Z-Score VV	32.61	28.39	33.21
Z-Score VH	2.79	11.30	17.32
Z-Score VV + VH	18.66	21.53	33.84
Z-Score VV − VH	41.57	16.37	8.56
Z-Score VV/VH	4.38	22.41	7.10

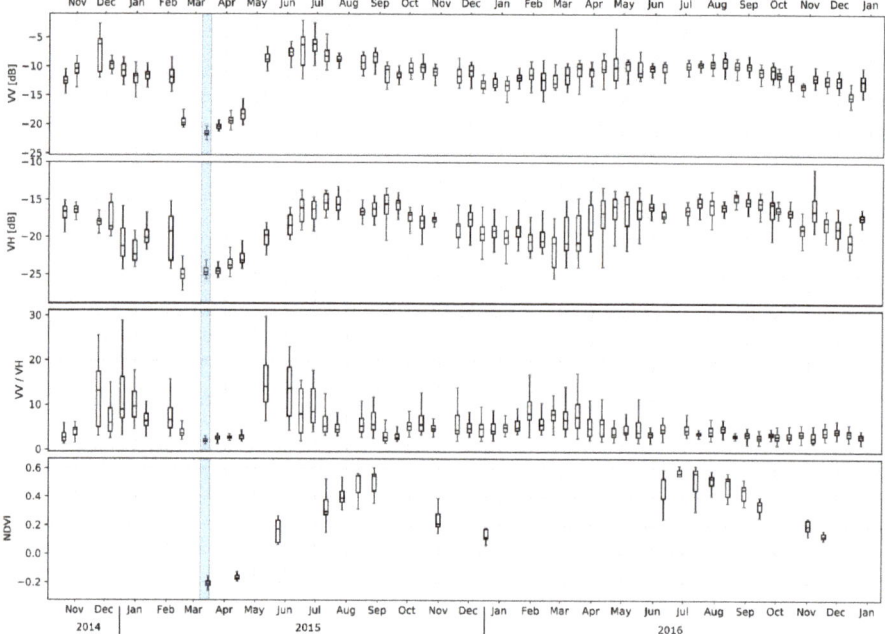

Figure 6. Multitemporal behavior of the backscatter intensity for TOW areas for VV (decibel (dB)), VH (dB), VV/VH (linear scale), and Normalized Difference Vegetation Index (NDVI) values in northern Greece/Turkey. The blue bars mark the analyzed date at the flood event.

For the study area in southern Greece/Turkey, the Z-Score VV − VH is the most relevant time series feature for the derivation of TFV (Table 5). This can be explained by the strong difference between the polarizations VV and VH (Figure 7). Thereby, VV polarization shows a strong increase at the flood date compared to the rest of the time series and VH polarization shows low or almost no increase at the flood date.

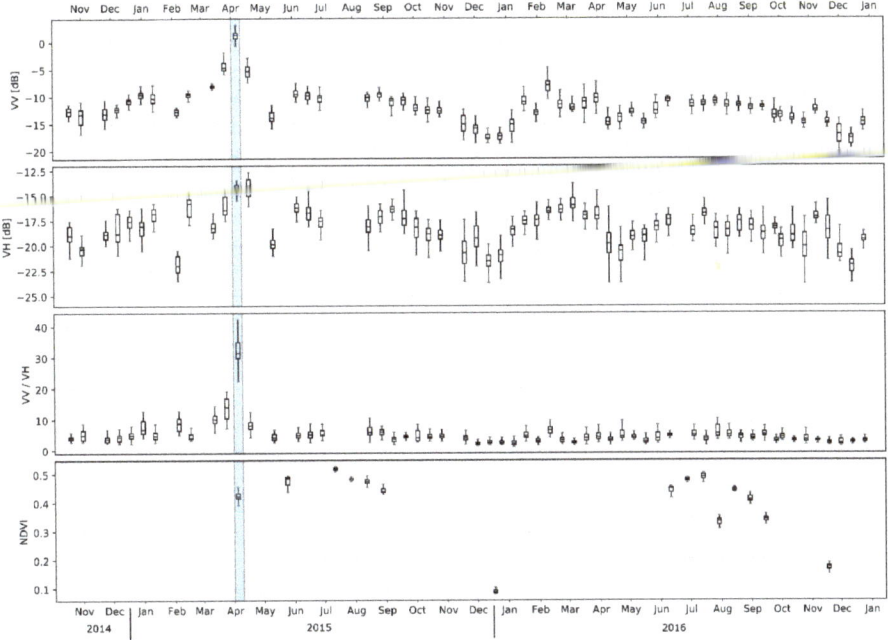

Figure 7. Multitemporal behavior of the backscatter intensity for TFV areas for VV (dB), VH (dB), VV/VH (linear scale), and NDVI values in southern Greece/Turkey. The blue bars mark the analyzed date at the flood event.

The different behavior of both polarizations can be explained by the different sensitivity of VV and VH to specific backscattering mechanisms, which caused an increased difference in the backscatter values between the two polarizations at the analyzed date of the flood. In comparison, the difference between the polarizations for the dry dates shows a smaller variability. The different sensitivity is on the one hand due to the fact that the double-bounce effect occurs more strongly in the VV polarization and leads to the backscatter. On the other hand, backscatter is in general lower for cross-polarization (HV or VH), as no ideal corner reflectors can be produced due to their depolarizing properties [5,58,59]. As a result, it is reported that the increase of the backscatter values by TFV can be more reliably detectable by co-polarized data than by cross-polarized data [60]. However, the combination of co- and cross-polarized data can improve the identification of TFV [26]. The NDVI values in Figure 7 confirm that the increase in VV polarization was not caused by seasonal or other changes in vegetation, but is flood-related, as no sudden increase in NDVI values could be identified at the flood date.

For the study area in northern Greece/Turkey, Z-Score VV is the most relevant feature (Table 5). As previously mentioned in Section 2.1, this study area is dominated by deciduous forest (Figure 2a), which was under leaf-on conditions at the date of the flood. It should be noted that the penetration of the forest canopy may be restricted by C-band and environmental parameters (e.g., aboveground biomass) [61,62]. Yu and Satschi [63] reported that the SAR backscatter values increase depending on the biomass and that the soil signal can no longer be detected if a certain degree of saturation has been reached. Thereby, the volume scattering from the canopy completely superimposes the contribution of the double bounce effects from the interaction between the water surface and vertical vegetation structure and TFV can no longer be identified. Therefore, there is hardly any signal difference between VV and VH at the flood date (Figure 8), which is confirmed by the low importance of Z-Score VV − VH. Compared to that the small increase in importance of the VV/VH can be explained by a small proportion of TFV in the forested areas due to limited penetration into the forest crown, as described above. In the case of northern Greece/Turkey the importance increases, because the outliers were eliminated by using the Z-Score VV/VH feature. For southern Greece/Turkey on the contrary, using the Z-Score VV/VH the outlier reduction caused also an elimination of extreme values caused by TFV. This led to a drop of the importance in VV/VH in comparison to the Z-Score VV − VH. Regarding the most relevant time series feature Z-Score VV for the northern Greece/Turkey study area, it is assumed that the backscatter in VV polarization is dominated by the contribution of the forest canopy and moreover the volume scattering represents the main scattering mechanism. Due to the sensitivity of the VV polarization for the double bounce effect, however, a slight increase in the backscattering values at the flood date was detected, which can indicate TFV (Figure 8). The NDVI values were derived from the LANDSAT 8 data and show that the slight increase of the backscatter values in both polarizations at the date of flood can be attributed to the flood event and not to any change in vegetation.

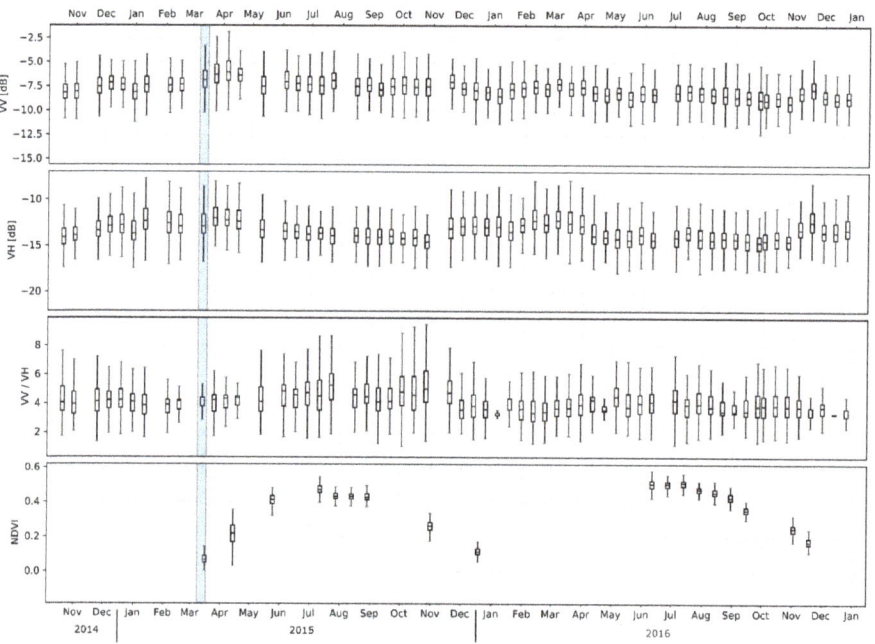

Figure 8. Multitemporal behavior of the backscatter intensity for TFV areas for VV (dB), VH (dB), VV/VH (linear scale), and NDVI values in northern Greece/Turkey. The blue bars mark the analyzed date at the flood event.

For the study area in China, Z-Score VV + VH represents the most relevant time series feature for the derivation of the TFV (Table 5). The time series feature indicates that in both polarizations an increase of the backscatter values has occurred at the analyzed date of the flood. This can be confirmed in Figure 9. Both polarizations increase at the analyzed flood date, whereas the stronger increase can be observed in VV polarization. The sum of the two polarizations intensifies the increase of the backscatter values at the same date. Although only a few cloud-free optical Sentinel-2 images are available that cover the study area in China for the analyzed time period, it can be observed that there is no increase in the NDVI values at the analyzed flood date (Figure 9). This leads to the deduction that the increase in SAR time series data is not related to phenological changes. Figure 5b shows a high-resolution Sentinel-2 image demonstrating the presence of the corresponding TFV area at the analyzed flood date.

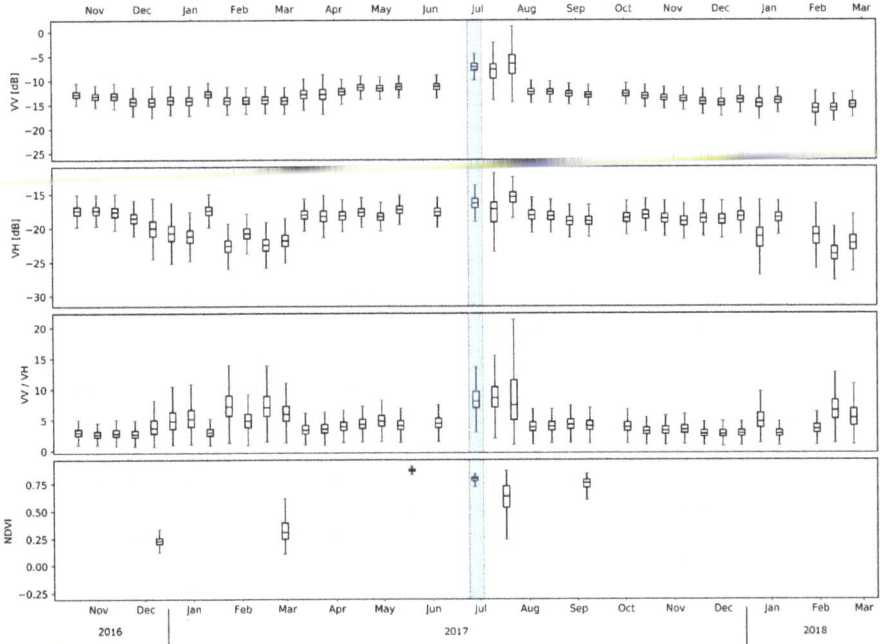

Figure 9. Multitemporal behavior of the backscatter intensity for TFV areas for VV (dB), VH (dB), VV/VH (linear scale), and NDVI values in China. The blue bars mark the analyzed date at the flood event.

Since the TFV type is high grassland and reed (Figure 5b) [64], which is similar to the structure of some crops (rice, wheat), it is assumed that the vegetation can be penetrated by the C-band data [10,65]. Therefore, it is expected that the increase in backscatter values will occur at the time of the flood in the VV polarization due to their sensitivity to the double impact effect. However, the increase in VH polarization cannot be confirmed by the depolarizing property mentioned above. Instead, the latter increase is a result of the combination of environmental parameters. The contribution of the soil under the vegetation plays a decisive role in this process. Due to of possible unevenness of the soil (microtopography) and the low water level, it is possible that a part of the soil surface beneath the vegetation was only partially flooded. In the flood-free areas underneath the vegetation, the soil moisture may have increased at the date of the flood compared to non-flooded conditions. Simultaneously, water could have accumulated in the sinks, which could also lead to standing water beneath vegetation in the same region. Due to different sensitivities of VV and VH to the respective conditions, the latter can dominate the respective polarization. The analyses from the literature confirm that an increase in soil moisture in both VV and VH causes an increase in the backscatter values and can be detected by the sensor depending on the biomass amount [66,67]. The TFV, which is represented by the double-bounce effect, can only be detected by VV polarization, as the depolarizing property of VH polarization does not allow sensitivity to double-bounce [68]. The mixture of these two cases by micro topographical differences explains the increase not only of backscatter values for VV, but also for VH and, thus, the relevance of Z-Score VV + VH for the derivation of TFV in the study area of China.

A comparison of the relevant time series features with the results of the study of Tsyganskaya et al., [8] shows that for the class TOW, Z-Score VV + VH is also one of the time series features with the highest contribution. For TFV, the most relevant time series characteristic (Z-Score VV − VH) is confirmed only for the study area in the southern area of interest of Greece/Turkey. In the other two example areas, the algorithm determined different relevant time series features due to different environmental conditions (e.g., vegetation type, topography, soil moisture) and their interaction.

Based on a previous analysis in Tsyganskaya et al., [8], the time series feature with the highest contribution can achieve a higher classification accuracy compared to the combination of all time series features. Therefore, a single time series feature with the highest contribution was used for the classification for deriving the desired classes. This identification step is performed automatically by the algorithm using training data. Moreover, these examples show that the use of both polarizations is relevant for the derivation of flooded areas.

3.2. Classification Results

The classification of the study area in southern Greece/Turkey was carried out based on the S-1 time series stack (October 19, 2014–December 25, 2016) containing the flood image (April 4, 2015). This classification was validated using the S-2-based reference mask (April 5, 2015). Figure 10a shows the pixel-based classification result of the time series approach, which comprises the classes POW, TOW, TFV, and DL, while Figure 10b shows the object-based classification result. For visual comparison, the validation mask is shown in Figure 1d. It is noticeable that the areas of the classes TOW, TFV, and DL contain more small structures of the other classes in the pixel-based classification compared to the object-based classification. The usage of objects reduces this noise; however, it also can lead to the loss of details and thus information. In general, the use of pixel- or object-based classification depends strongly on the kind of the landscape under investigation. In the case of flooding in vegetated areas, a single, isolated pixel could contain water ponds and so differ from its neighboring pixels. Due to limited visibility in optical data, these areas could not be identified during the generation of the validation mask.

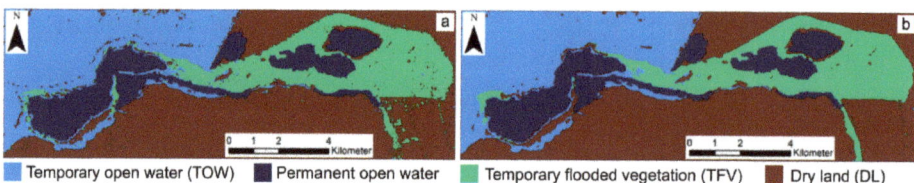

Figure 10. Pixel-based classification result (**a**), object-based classification result (**b**) for the study area in southern Greece/Turkey.

For the study area in northern Greece/Turkey, the classification is based on the S-1 time series stack (October 19, 2014–December 25, 2016), with the flood image being acquired on March 12, 2015. Validation of this classification was performed using the S-2-based reference mask (March 15, 2015). Figure 11a shows the pixel-based classification result for the study area of northern Greece/Turkey, which comprises the classes POW, TOW, TFV, and DL, while Figure 11b shows the object-based classification result. Compared to the validation mask (Figure 1c), both pixel- and object-based classification show only a low representativity of the class TFV. This can be explained by the low penetration of the vegetation (deciduous forest (Figure 2a)) by the C-band data, which is reduced in this case [8]. In addition, occasional TFV areas in the classification can be identified in DL of the validation mask, especially in the pixel-based classification. These could be small areas of water (ponds or temporary water accumulations). In combination with vertical structures of the agricultural areas or grassland, the double bounce effect can also be produced resulting in an increase in the backscatter values at the date of the flood. These areas were also classified as TFV, though they could not, however, be identified during the generation of the validation mask due to limited visibility in optical data.

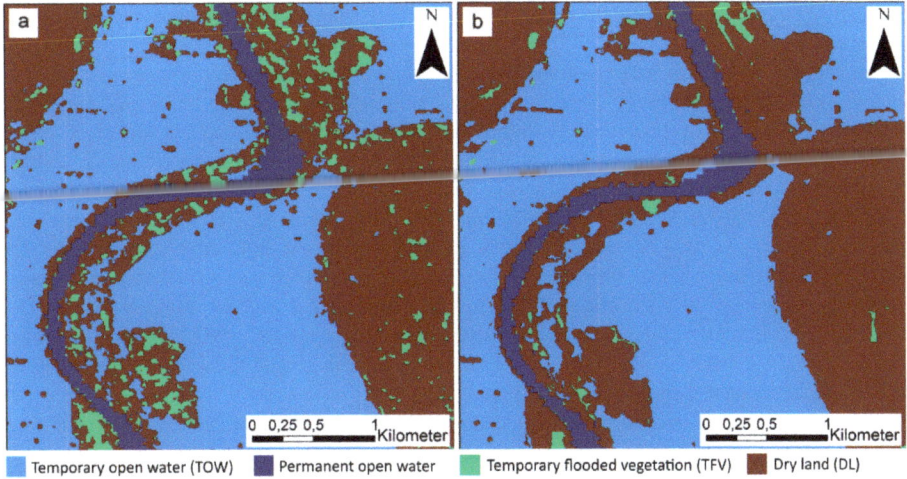

Figure 11. Pixel-based classification result (**a**) and object-based classification result (**b**) for the study area in northern Greece/Turkey.

The classification for the study area in China was performed using the S-1 time series stack (October 19, 2016–February 23, 2018). The analyzed date of the flood is June 28, 2017. The validation for this classification was performed using the S-2-based reference mask (June 27, 2017). Figure 12a, b show the pixel- and object-based classification results with POW, TOW, TFV, and DL classes. The validation mask used is shown in Figure 4c.

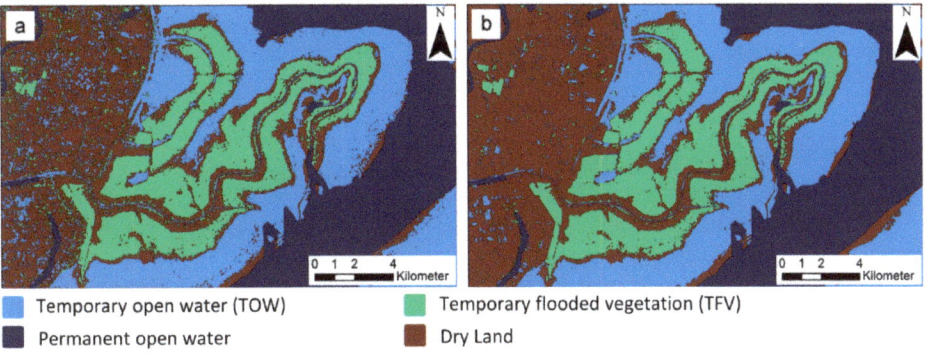

Figure 12. Pixel-based classification result (**a**), object-based classification result (**b**) and validation mask (**c**) for the study area in China.

The comparison between the classification results and the validation mask reveals the occurrence of TOW in DL. DL also occurs between the permanent water surfaces and the TOW areas. On the one hand, the confusion between these two classes can arise due to small areas of water occurring in the agricultural areas. On the other hand, the duration of a flood event can lead to an accumulation of S-1 images, containing the inundation and a decrease of images without flooding. This can cause a higher fluctuation range and a lower mean value when generating the time series features, which in particular can lead to confusion with DL, as the areas between permanent water and TOW demonstrate. The combination of these two conditions makes the above-mentioned confusions between the classes possible. Therefore, the use of a time series with at least one vegetation cycle is recommended, as the statistical range of the backscatter values of the vegetation stages is detected and thus sufficiently represented [8].

The accuracy values for the pixel-based and object-based classifications for the respective study areas are shown in Table 6. The OA for all study areas ranges between 80% and 87%. In comparison, in [8] the authors achieved a similar or even slightly lower OA for the pixel-based (75%) and object-based (82%) classification. The accuracy assessment for the study area in northern Greece/Turkey shows high values for the TOW class but strong misclassifications for TFV (PA: 28% for pixel-based and 6% for object-based) and DL (UA: 57% for pixel-based and 53% for object-based). This indicates a significant confusion between these two classes and can be explained by different environmental conditions (see Section 3.1). The accuracy values show no significant difference between pixel-based and object-based classification, except for the PA values of TFV. When applying object-based classification, it should be noted that the small-scale flood areas could be generalized by merging the pixels to objects. However, object-based classification can help to exclude pixels which can be confused with a flood-related increase in values due to dominant environmental conditions, such as soil moisture, topography, or surface roughness, which can cause a small-scale increase or decrease in the backscatter. In addition, objects are less susceptible to speckle noise than pixels [29].

Table 6. Accuracy assessment for the pixel- and object-based classification for the individual study areas. Overall accuracy (OA) in %, Producer accuracy (PA) in %, User accuracy (UA) in %, and Kappa index (K).

	Southern Greece/Turkey (%)		Northern Greece/Turkey (%)		China (%)	
	Pixel-based	Object-based	Pixel-based	Object-based	Pixel-based	Object-based
DL—UA	76.64	77.62	56.65	53.25	85.26	87.66
TOW—UA	86.53	86.87	97.58	98.23	79.73	85.65
TFV—UA	90.37	92.17	76.83	73.86	78.28	83.01
DL—PA	77.86	79.61	91.67	96.60	69.89	78.95
TOW—PA	85.94	86.38	91.02	91.67	90.47	91.14
TFV—PA	89.52	90.09	28.18	6.32	90.09	91.23
OA	84.26	85.17	81.47	79.59	81.37	85.84
Kappa	0.76	0.78	0.66	0.63	0.71	0.78

3.3. Time Series Features—Transferability Analysis

For the transferability analysis of the classification approach, several pixel-based and object-based classification runs were carried out for each study area and its subareas using the individual time series features. For each run, the most relevant time series feature (Z-Score VV + VH) was used for the derivation of TOW (Section 3.1). Simultaneously, each time series feature was used in each study area to derive the TFV, so that 20 pixel-based and 20 object-based classification products could be generated. For each result, an accuracy assessment was performed. Since the analysis of the different time series features refers to the TFV, the PA and UA of TFV were selected as indicators to represent the classification accuracy of the TFV. The accuracy values PA and UA for the individual study areas and the five analyzed time series features for pixel-based classification are shown in Table 7 and for object-based classification in Table 8. The relevant time series features that were identified by the Random Forest Algorithm (Section 3.1) are highlighted in dark green for the respective study areas for the PA and UA values. These tables show that different time series features are relevant for the individual study areas. The low PA values seen for northern Greece in both tables can be explained by the specific TFV type that dominates this study area, namely flooded forest areas. In this case, the penetration of the forest crown is limited by C-band data, whereby the water under the vegetation can only be partially detected.

Table 7. The accuracy values PA and UA of the pixel-based classification for the individual study areas as a function of the five analyzed time series features. The relevant time series features for the individual study areas, which were identified by the Random Forest Algorithm (Section 3.1) and corresponding PA and UA values are highlighted in dark green. Optimum of statistical variables (mean value of UA / PA and coefficient of variables) for time series features is highlighted in light green.

	Z-Score VV		Z-Score VH		Z-Score VV + VH		Z-Score VV − VH		Z-Score VV/VH	
	UA	PA	UA	PA	UA	PA	UA	PA	UA	PA
Namibia	88.72	78.43	47.11	90.52	85.64	82.38	92.62	69.96	67.90	86.32
China	82.26	92.65	51.26	99.30	78.28	90.09	52.37	99.24	44.46	99.50
North Greece	76.83	28.18	50.34	0.81	65.85	3.26	37.10	4.71	68.58	56.72
South Greece	96.10	77.17	35.25	42.22	94.89	53.17	90.37	89.52	60.85	93.48
Mean values of UA and PA	77.54		52.10		69.18		65.74		72.23	
Coefficient of variance	0.22		0.41		0.36		0.47		0.18	

Table 8. The accuracy values PA and UA of the object-based classification for the individual study areas as a function of the five analyzed time series features. The relevant time series features for the individual study areas, which were identified by the Random Forest Algorithm (Section 3.1) and corresponding PA and UA values are highlighted in dark green. Optimum of statistical variables (mean value of UA / PA and coefficient of variables) for time series features is highlighted in light green.

	Z-Score VV		Z-Score VH		Z-Score VV + VH		Z-Score VV − VH		Z-Score VV/VH	
	UA	PA	UA	PA	UA	PA	UA	PA	UA	PA
Namibia	84.05	80.71	49.26	17.47	84.37	5.99	54.93	89.29	76.1	91.2
China	82.38	92.65	62.51	92.46	83.01	91.23	66.68	97.06	42.86	97.86
North Greece	73.86	6.32	50.35	0.82	65.99	3.29	48.61	2.92	68.75	57.22
South Greece	97.05	87.81	26.90	31.64	98.52	75.19	92.17	90.09	65.20	92.74
Mean values of UA and PA	75.57		41.43		63.44		67.72		73.99	
Coefficient of variance	0.31		0.62		0.52		0.40		0.19	

As described in the methodology, two quantities, mean value and CV, were derived based on the UA and PA values to provide a foundation for the identification of a single robust time series feature for the derivation of TFV types and for all study areas. If a time series feature reaches a high mean value compared to other time series features and shows at the same time the smallest CV, it is relevant for all study areas. Considering Tables 7 and 8, it is not apparent which time series feature would be suitable since the two statistical variables achieve the optimum at different time series features (highlighted in light green). The Z-Score VV shows the highest mean value (77.54%) for pixel-based classification and (75.57%) for object-based classification compared to other time series features, but Z-Score VV/VH shows the lowest CV value (0.18) for pixel-based and (0.19) for object-based.

For a distinct identification of the most relevant time series feature, the two statistical variables have to be considered simultaneously. In Figure 13, the two quantities are plotted on two different axes. The larger the mean value and the smaller the CV, the more suitable is a time series feature for the extraction of TFV for all study areas. The lower right corner of the diagram is thus the point that represents the optimum between both statistical quantities. The closer a time series feature lies to this optimum, the more relevant is the time series feature for all study areas.

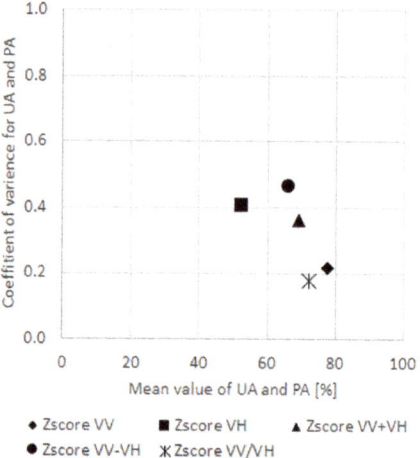

Figure 13. Comparison of pixel-based time series features as a function of mean value and coefficient of variation. The statistical values (mean value and coefficient of variance) were calculated on the basis of the PA and UA accuracy values.

Compared to other time series features, Z-Score VV is the shortest regarding the distance between the optimum and the time series feature for pixel-based classification (Figure 13). Thus, this time series feature seems to be the most relevant for the operational derivation of the class TFV, when using pixels as a foundation. The relevance of Z-Score VV is due to the fact that VV polarization is more influenced by the double-bounce effect indicating the presence of TFV. In comparison, VH is more influenced by different environment conditions and may change for different TFV types and study areas (Section 3.1). The distances of the Z-Score VV/VH to the optimal point differ from the distance between the Z-Score VV and optimal point only slightly. This shows that VH polarization also has an influence on the classification results and can be important in combination with VV polarization provided that the environmental conditions in the analyzed study area are known and characterized.

This relevance is also shown in Figure 14, where the relevant feature based on objects is the Z-Score VV/VH, which contains both polarizations. Based on the findings in Section 3.1, which show that the increase in the VV time series at the date of the flood for TFV is more significant compared to the VH time series because the VH signal is influenced more by environmental conditions than by TFV, the Z-Score VV feature based on pixel elements is recommended for operational use. In addition, the mean value of UA and PA for the pixel-based classification for the Z-Score VV time series feature is highest (77.54%) compared to all other mean values of UA and PA. The combination of Z-Score VV + VH and Z-Score VV/VH based on pixel elements is recommended for operative use for extraction of the TOW and TFV, respectively. When no feature importance has to be calculated during the classification and the pre-processing of the S-1 data [8] has already taken place, the classification process takes between 1 and 5 minutes, depending on the available computing performance and the extent of the study area. The maximum extent that was analyzed is one-third of S-1 GRD images. By specifying the time series features the user, interaction is omitted and apart from the initialization of the algorithm the classification can be performed automatically. For the optimization and extension of the approach regarding different TFV types, further research in other study areas will be beneficial.

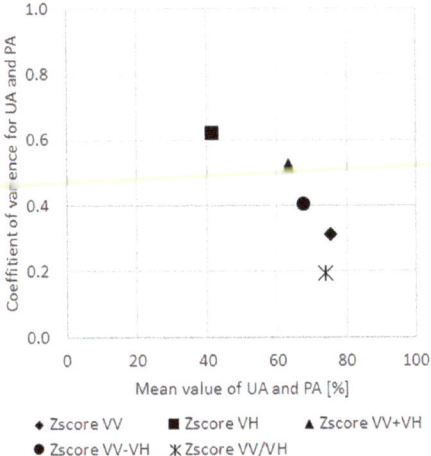

Figure 14. Comparison of object-based time series features as a function of mean value and coefficient of variation. The statistical values (mean value and coefficient of variance) were calculated on the basis of the PA and UA accuracy values.

4. Conclusions

The results of the study areas in northern and southern Greece/Turkey and in China show the potential of the method presented in Tsyganskaya et al., [8] using S-1 time series data to extract the total flood extent considering both temporary open water (TOW) and temporary flooded vegetation (TFV). In particular, different types of TFV were analyzed. The individual flood extent was determined by time series features, which represent the characteristics of both classes.

This study confirms the results of the study of Tsyganskaya et al., [8], showing that the Z-Score VV+VH, which was derived based on the characteristics mentioned above, is the most relevant time series feature for the extraction of the TOW areas in all study areas. For the derivation of TFV, three different time series features were determined by the algorithm for individual study areas. This can be explained by the complex structure of the vegetation, the various analyzed vegetation types, and the dependency of the TFV on environmental conditions (e.g., vegetation type, soil moisture, topography), which differ in all study areas. Nevertheless, in the results for all study areas OA values reached between 80% and 87%, which were slightly higher even compared to the results in [8]. For the study area in northern Greece, the TFV areas could only partially be derived based on the C-band data due to the presence of forested vegetation, which reduces the penetration of the SAR signal. In both other study areas, TFV in high grassland/reed was successfully classified.

The potential of the time series approach for transferability as a prerequisite for operational use was analyzed by comparing time series features with respect to their suitability for the derivation of TFV for all study areas simultaneously. For the comparison, statistical quantities were derived from the classification accuracies PA and UA of the TFV. For TFV, Z-Score VV, or Z-Score VV/VH appear to be the most relevant time series feature for pixel-based and object-based classification, respectively. Nevertheless, Z-Score VV is recommended for operational use based on pixel elements for the extraction of TFV as this allows improved differentiation to non-flood dates due to the more significant increase in the VV time series for TFV at the date of the flood compared to the VH time series. The Z-Score VV + VH is recommended for the extraction of TOW surfaces, because of the unanimity regarding the feature importance calculation for all study areas (Section 3.1). Since the time series features have been defined and no feature importance needs to be calculated, the user interaction is reduced to the initialization of the classification process and the computation time decreases.

Overall, the approach can be used for irregularly occurring flood events, as well as for irregularly acquired S-1 images. It is flexible for individual applications depending on the vegetation and takes account of the seasonal changes by the use of multitemporal data. The presented SAR time series approach lays the cornerstone for automatic flood detection on a global scale, allowing the detection of the entire flood extent by supplementing the TOW with TFV areas.

Author Contributions: The concept of this study was developed by V.T., P.M. and S.M., while V.T. was responsible for the implementation, which included the preparation of data, generation and validation of flood classification maps, and methodological comparison of relevant features and their discussion.

Funding: This work is funded by the Federal Ministry for Economic Affairs and Energy (BMWi), grant number 50 EE1338.

Acknowledgments: The WorldView-2 imagery was kindly provided by European Space Imaging Ltd. (EUSI). The RapidEye imagery was kindly provided by Planet Labs Inc.

Conflicts of Interest: The authors declare no conflict of interest.

References

1. The International Disaster-Emergency Events Database (EMDAT). *OFDA/CRED International Disaster Database*; Université catholique de Louvain: Brussels, Belgium, 2019; Available online: https://www.emdat.be (accessed on 1 August 2019).
2. Centre for Research on the Epidemiology of Disasters (CRED); The United Nations Office for Disaster Risk Reduction (UNISDR). 2018 Review of Disaster Events. 2019. Available online: https://www.emdat.be/publications (accessed on 31 August 2019).
3. Klemas, V. Remote Sensing of Floods and Flood-Prone Areas: An Overview. *J. Coast. Res.* **2014**, *31*, 1005–1013. [CrossRef]
4. Cazals, C.; Rapinel, S.; Frison, P.-L.; Bonis, A.; Mercier, G.; Mallet, C.; Corgne, S.; Rudant, J.-P. Mapping and Characterization of Hydrological Dynamics in a Coastal Marsh Using High Temporal Resolution Sentinel-1A Images. *Remote Sens.* **2016**, *8*, 570. [CrossRef]
5. Martinis, S.; Rieke, C. Backscatter Analysis Using Multi-Temporal and Multi-Frequency SAR Data in the Context of Flood Mapping at River Saale, Germany. *Remote Sens.* **2015**, *7*, 7732–7752. [CrossRef]
6. Pulvirenti, L.; Pierdicca, N.; Chini, M. Analysis of Cosmo-SkyMed observations of the 2008 flood in Myanmar. *Ital. J. Remote Sens.* **2010**, *42*, 79–90. [CrossRef]
7. Brisco, B.; Shelat, Y.; Murnaghan, K.; Montgomery, J.; Fuss, C.; Olthof, I.; Hopkinson, C.; Deschamps, A.; Poncos, V. Evaluation of C-Band SAR for Identification of Flooded Vegetation in Emergency Response Products. *Can. J. Remote Sens.* **2019**. [CrossRef]
8. Tsyganskaya, V.; Martinis, S.; Marzahn, P.; Ludwig, R. Detection of Temporary Flooded Vegetation Using Sentinel-1 Time Series Data. *Remote Sens.* **2018**, *10*, 1286. [CrossRef]
9. Moser, L.; Schmitt, A.; Wendleder, A. Automated Wetland Delineation from Multi-Frequency and Multi-Polarized SAR Images in High Temporal and Spatial Resolution. *ISPRS Ann. Photogramm. Remote Sens. Spat. Inf. Sci.* **2016**, *3*, 57–64. [CrossRef]
10. Pulvirenti, L.; Pierdicca, N.; Chini, M.; Guerriero, L. Monitoring Flood Evolution in Vegetated Areas Using COSMO-SkyMed Data: The Tuscany 2009 Case Study. *IEEE J. Sel. Top. Appl. Earth Obs. Remote Sens.* **2012**, *6*, 1807–1816. [CrossRef]
11. Betbeder, J.; Rapinel, S.; Corpetti, T.; Pottier, E.; Corgne, S.; Hubert-Moy, L. Multitemporal Classification of TerraSAR-X Data for Wetland Vegetation Mapping. *J. Appl. Remote Sens.* **2014**, *8*, 83648. [CrossRef]
12. Chapman, B.; McDonald, K.; Shimada, M.; Rosenqvist, A.; Schroeder, R.; Hess, L. Mapping Regional Inundation with Spaceborne L-Band SAR. *Remote Sens.* **2015**, *7*, 5440–5470. [CrossRef]
13. Evans, T.L.; Costa, M.; Tomas, W.M.; Camilo, A.R. Large-Scale Habitat Mapping of the Brazilian Pantanal Wetland. A synthetic aperture radar approach. *Remote Sens. Environ.* **2014**, *155*, 89–108. [CrossRef]
14. Hess, L. Dual-Season Mapping of Wetland Inundation and Vegetation for the Central Amazon Basin. *Remote Sens. Environ.* **2003**, *87*, 404–428. [CrossRef]

15. Lang, M.W.; Kasischke, E.S.; Prince, S.D.; Pittman, K.W. Assessment of C-band synthetic aperture radar data for mapping and monitoring Coastal Plain forested wetlands in the Mid-Atlantic Region, USA. *Remote Sens. Environ.* **2008**, *112*, 4120–4130. [CrossRef]
16. Chini, M.; Papastergios, A.; Pulvirenti, L.; Pierdicca, N.; Matgen, P.; Parcharidis, I. SAR coherence and polarimetric information for improving flood mapping. In Proceedings of the 2016 IEEE International Geoscience & Remote Sensing Symposium, Beijing, China, 10–15 July 2016; pp. 7577–7580.
17. Morandeira, N.; Grings, F.; Facchinetti, C.; Kandus, P. Mapping Plant Functional Types in Floodplain Wetlands. An Analysis of C-Band Polarimetric SAR Data from RADARSAT-2. *Remote Sens.* **2016**, *8*, 174. [CrossRef]
18. Touzi, R.; Deschamps, A.; Rother, G. Wetland Characterization using Polarimetric RADARSAT-2 Capability. *Can. J. Remote Sens.* **2007**, *33*, S56–S67. [CrossRef]
19. Pulvirenti, L.; Chini, M.; Pierdicca, N.; Boni, G. Use of SAR Data for Detecting Floodwater in Urban and Agricultural Areas: The Role of the Interferometric Coherence. *IEEE Trans. Geosci. Remote Sens.* **2016**, *54*, 1532–1544. [CrossRef]
20. Baghdadi, N.; Bernier, M.; Gauthier, R.; Neeson, I. Evaluation of C-band SAR Data for Wetlands Mapping. *Int. J. Remote Sens.* **2001**, *22*, 71–88. [CrossRef]
21. Gallant, A.; Kaya, S.; White, L.; Brisco, B.; Roth, M.; Sadinski, W.; Rover, J. Detecting Emergence, Growth, and Senescence of Wetland Vegetation with Polarimetric Synthetic Aperture Radar (SAR) Data. *Water* **2014**, *6*, 694–722. [CrossRef]
22. White, L.; Brisco, B.; Pregitzer, M.; Tedford, B.; Boychuk, L. RADARSAT-2 Beam Mode Selection for Surface Water and Flooded Vegetation Mapping. *Can. J. Remote Sens.* **2014**, *40*, 135–151.
23. Brisco, B.; Schmitt, A.; Murnaghan, K.; Kaya, S.; Roth, A. SAR Polarimetric Change Detection for Flooded Vegetation. *Int. J. Digit. Earth* **2011**, *6*, 103–114. [CrossRef]
24. De Grandi, G.F.; Mayaux, P.; Malingreau, J.P.; Rosenqvist, A.; Saatchi, S.; Simard, M. New Perspectives on Global Ecosystems from Wide-Area Radar Mosaics. Flooded forest mapping in the tropics. *Int. J. Remote Sens.* **2010**, *21*, 1235–1249. [CrossRef]
25. Dabboor, M.; White, L.; Brisco, B.; Charbonneau, F. Change Detection with Compact Polarimetric SAR for Monitoring Wetlands. *Can. J. Remote Sens.* **2015**, *41*, 408–417. [CrossRef]
26. Zhao, L.; Yang, J.; Li, P.; Zhang, L. Seasonal Inundation Monitoring and Vegetation Pattern Mapping of the Erguna Floodplain by Means of a RADARSAT-2 Fully Polarimetric Time Series. *Remote Sens. Environ.* **2014**, *152*, 426–440. [CrossRef]
27. Koch, M.; Schmid, T.; Reyes, M.; Gumuzzio, J. Evaluating Full Polarimetric C- and L-Band Data for Mapping Wetland Conditions in a Semi-Arid Environment in Central Spain. *IEEE J. Sel. Top. Appl. Earth Obs. Remote Sens.* **2012**, *5*, 1033–1044. [CrossRef]
28. Lee, J.-S.; Pottier, E. *Polarimetric Radar Imaging: From Basics to Applicationsvol*; CRC Press: Boca Raton, FL, USA, 2009; p. 142.
29. Tsyganskaya, V.; Martinis, S.; Marzahn, P.; Ludwig, R. SAR-based Detection of Flooded Vegetation–A Review of Characteristics and Approaches. *Int. J. Remote Sens.* **2018**, *39*, 2255–2293. [CrossRef]
30. Chini, M.; Pulvirenti, L.; Pierdicca, N. Analysis and Interpretation of the COSMO-SkyMed Observations of the 2011 Japan Tsunami. *IEEE Geosci. Remote Sens. Lett.* **2012**, *9*, 467–471. [CrossRef]
31. Long, S.; Fatoyinbo, T.E.; Policelli, F. Flood Extent Mapping for Namibia using Change Detection and Thresholding with SAR. *Environ. Res. Lett.* **2014**, *9*, 035002. [CrossRef]
32. Voormansik, K.; Praks, J.; Antropov, O.; Jagomagi, J.; Zalite, K. Flood Mapping with TerraSAR-X in Forested Regions in Estonia. *IEEE J. Sel. Top. Appl. Earth Obs. Remote Sens.* **2014**, *7*, 562–577. [CrossRef]
33. Horritt, M.S.; Mason, D.C.; Luckman, A.J. Flood Boundary Delineation from Synthetic Aperture Radar Imagery Using a Statistical Active Contour Model. *Int. J. Remote Sens.* **2001**, *22*, 2489–2507. [CrossRef]
34. Martinis, S.; Twele, A. A Hierarchical Spatio-Temporal Markov Model for Improved Flood Mapping Using Multi-Temporal X-Band SAR Data. *Remote Sens.* **2010**, *2*, 2240–2258. [CrossRef]
35. Chen, Y.; He, X.; Wang, J.; Xiao, R. The Influence of Polarimetric Parameters and an Object-Based Approach on Land Cover Classification in Coastal Wetlands. *Remote Sens.* **2014**, *6*, 12575–12592. [CrossRef]
36. Plank, S.; Jüssi, M.; Martinis, S.; Twele, A. Mapping of Flooded Vegetation by Means of Polarimetric Sentinel-1 and ALOS-2/PALSAR-2 imagery. *Int. J. Remote Sens.* **2017**, *38*, 3831–3850. [CrossRef]

37. Karszenbaum, H.; Kandus, P.; Martinez, J.M.; Le Toan, T.; Tiffenberg, J.; Parmuchi, G. *Radarsat SAR Backscattering Characteristics of the Parana River Delta Wetland, Argentina*; ESA Publication: Auckland, New Zealand, 2000.
38. Pierdicca, N.; Chini, M.; Pulvirenti, L.; Macina, F. Integrating Physical and Topographic Information into a Fuzzy Scheme to Map Flooded Area by SAR. *Sensors* **2008**, *8*, 4151–4164. [CrossRef] [PubMed]
39. Pulvirenti, L.; Pierdicca, N.; Chini, M.; Guerriero, L. An Algorithm for Operational Flood Mapping from Synthetic Aperture Radar (SAR) Data using Fuzzy Logic. *Nat. Hazards Earth Syst. Sci.* **2011**, *2*, 529–540. [CrossRef]
40. Bouvet, A.; Le Toan, T. Use of ENVISAT/ASAR wide-swath data for timely rice fields mapping in the Mekong River Delta. *Remote Sens. Environ.* **2011**, *115*, 1090–1101. [CrossRef]
41. Martinez, J.; Le Toan, T. Mapping of Flood Dynamics and Spatial Distribution of Vegetation in the Amazon Floodplain using Multitemporal SAR Data. *Remote Sens. Environ.* **2007**, *108*, 209–223. [CrossRef]
42. Hess, L.L.; Melack, J.M.; Affonso, A.G.; Barbosa, C.; Gastil-Buhl, M.; Novo, E.M.L.M. Wetlands of the Lowland Amazon Basin. Extent, Vegetative Cover, and Dual-season Inundated Area as Mapped with JERS-1 Synthetic Aperture Radar. *Wetlands* **2015**, *35*, 745–756. [CrossRef]
43. Schlaffer, S.; Chini, M.; Dettmering, D.; Wagner, W. Mapping Wetlands in Zambia Using Seasonal Backscatter Signatures Derived from ENVISAT ASAR Time Series. *Remote Sens.* **2016**, *8*, 402. [CrossRef]
44. Ferreira-Ferreira, J.; Silva, T.S.F.; Streher, A.S.; Affonso, A.G.; de Almeida Furtado, L.F.; Forsberg, B.R.; de Moraes Novo, E.M.L. Combining ALOS/PALSAR derived vegetation structure and inundation patterns to characterize major vegetation types in the Mamirauá Sustainable Development Reserve, Central Amazon floodplain, Brazil. *Wetlands Ecol. Manag.* **2015**, *23*, 41–59. [CrossRef]
45. Lee, H.; Yuan, T.; Jung, H.C.; Beighley, E. Mapping wetland water depths over the central Congo Basin using PALSAR ScanSAR, Envisat altimetry, and MODIS VCF data. *Remote Sens. Environ.* **2015**, *159*, 70–79. [CrossRef]
46. Li, J.; Chen, W. A rule-based method for mapping Canada's wetlands using optical, radar and DEM data. *Int. J. Remote Sens.* **2005**, *22*, 5051–5069. [CrossRef]
47. Marti-Cardona, B.; Dolz-Ripolles, J.; Lopez-Martinez, C. Wetland inundation monitoring by the synergistic use of ENVISAT/ASAR imagery and ancilliary spatial data. *Remote Sens. Environ.* **2013**, *139*, 171–184. [CrossRef]
48. Bourgeau-Chavez, L.; Lee, Y.; Battaglia, M.; Endres, S.; Laubach, Z.; Scarbrough, K. Identification of Woodland Vernal Pools with Seasonal Change PALSAR Data for Habitat Conservation. *Remote Sens.* **2016**, *8*, 490. [CrossRef]
49. Zhang, M.; Li, Z.; Tian, B.; Zhou, J.; Zeng, J. A Method for Monitoring Hydrological Conditions Beneath Herbaceous Wetlands Using Multi-temporal ALOS PALSAR Coherence Data. *Int. Arch. Photogramm. Remote Sens. Spat. Inf. Sci.* **2015**, *6*, 221–226. [CrossRef]
50. Evans, T.L.; Costa, M. Landcover classification of the Lower Nhecolândia subregion of the Brazilian Pantanal Wetlands using ALOS/PALSAR, RADARSAT-2 and ENVISAT/ASAR imagery. *Remote Sens. Environ.* **2013**, *128*, 118–137. [CrossRef]
51. Grings, F.M.; Ferrazzoli, P.; Karszenbaum, H.; Salvia, M.; Kandus, P.; Jacobo-Berlles, J.C.; Perna, P. Model investigation about the potential of C band SAR in herbaceous wetlands flood monitoring. *Int. J. Remote Sens.* **2008**, *29*, 5361–5372. [CrossRef]
52. Lang, M.W.; Kasischke, E.S. Using C-Band Synthetic Aperture Radar Data to Monitor Forested Wetland Hydrology in Maryland's Coastal Plain, USA. *IEEE Trans. Geosci. Remote Sens.* **2008**, *46*, 535–546. [CrossRef]
53. Farr, T.G.; Rosen, P.A.; Caro, E.; Crippen, R.; Duren, R.; Hensley, S.; Kobrick, M.; Paller, M.; Rodriguez, E.; Roth, L.; et al. The Shuttle Radar Topography Mission. *Rev. Geophys.* **2007**, *45*, 1485. [CrossRef]
54. Esch, T.; Taubenböck, H.; Roth, A.; Heldens, W.; Felbier, A.; Thiel, M.; Schmidt, M.; Müller, A.; Dech, S. TanDEM-X Mission—New Perspectives for the Inventory and Monitoring of Global Settlement Patterns. *J. Appl. Remote Sens.* **2012**, *6*, 061702. [CrossRef]
55. Rennó, C.D.; Nobre, A.D.; Cuartas, L.A.; Soares, J.V.; Hodnett, M.G.; Tomasella, J.; Waterloo, M.J. HAND, a New Terrain Descriptor Using SRTM-DEM: Mapping terra-firme rainforest environments in Amazonia. *Remote Sens. Environ.* **2008**, *112*, 3469–3481. [CrossRef]
56. Twele, A.; Cao, W.; Plank, S.; Martinis, S. Sentinel-1-based Flood Mapping: A fully Automated Processing Chain. *Int. J. Remote Sens.* **2016**, *37*, 2990–3004. [CrossRef]

57. Richards, J.A. *Remote Sensing Digital Image Analysis*; Springer: Berlin, Germany, 2013.
58. Lewis, F.M.; Henderson, A.J. *Principles and Applications of Imaging Radar: Manual of Remote Sensingvol*; Wiley: New York, NY, USA, 1998; Volume 2.
59. Marti-Cardona, B.; Lopez-Martinez, C.; Dolz-Ripolles, J.; Bladè-Castellet, E. ASAR Polarimetric, Multi-Incidence Angle and Multitemporal Characterization of Doñana Wetlands for Flood Extent Monitoring. *Remote Sens. Environ.* **2010**, *114*, 2802–2815. [CrossRef]
60. Hess, L.L.; Melack, J.M.; Simonett, D.O. Radar Detection of Flooding Beneath the Forest Canopy: A review. *Int. J. Remote Sens.* **1990**, *11*, 1313–1325. [CrossRef]
61. Sang, H.; Zhang, J.; Lin, H.; Zhai, L. Multi-Polarization ASAR Backscattering from Herbaceous Wetlands in Poyang Lake Region, China. *Remote Sens.* **2014**, *6*, 4621–4646. [CrossRef]
62. Woodhouse, I.H. *Introduction to Microwave Remote Sensing*; CRC Press Taylor & Francis: Boca Raton, FL, USA, 2006.
63. Yu, Y.; Saatchi, S. Sensitivity of L-Band SAR Backscatter to Aboveground Biomass of Global Forests. *Remote Sens.* **2016**, *8*, 522. [CrossRef]
64. Hu, J.-Y.; Xie, Y.-H.; Tang, Y.; Li, F.; Zou, Y.-A. Changes of Vegetation Distribution in the East Dongting Lake After the Operation of the Three Gorges Dam, China. *Front. Plant Sci.* **2018**, *9*, 582. [CrossRef] [PubMed]
65. Ulaby, F.T.; Long, D.G. *Microwave Radar and Radiometric Remote Sensing*; Artech House: Norwood, Switzerland, 2015.
66. Bousbih, S.; Zribi, M.; Lili-Chabaane, Z.; Baghdadi, N.; El Hajj, M.; Gao, Q.; Mougenot, B. Potential of Sentinel-1 Radar Data for the Assessment of Soil and Cereal Cover Parameters. *Sensors* **2017**, *17*, 2617. [CrossRef]
67. Kasischke, E.S.; Smith, K.B.; Bourgeau-Chavez, L.L.; Romanowicz, E.A.; Brunzell, S.; Richardson, C.J. Effects of Seasonal Hydrologic Patterns in South Florida Wetlands on Radar Backscatter Measured from ERS-2 SAR Imagery. *Remote Sens. Environ.* **2003**, *88*, 423–441. [CrossRef]
68. Kwoun, O.; Lu, Z. Multi-temporal RADARSAT-1 and ERS Backscattering Signatures of Coastal Wetlands in Southeastern Louisiana. *Photogramm. Eng. Remote Sens.* **2009**, *75*, 607–617. [CrossRef]

© 2019 by the authors. Licensee MDPI, Basel, Switzerland. This article is an open access article distributed under the terms and conditions of the Creative Commons Attribution (CC BY) license (http://creativecommons.org/licenses/by/4.0/).

Article

Integrating C- and L-Band SAR Imagery for Detailed Flood Monitoring of Remote Vegetated Areas

Alberto Refice [1,*], Marina Zingaro [2], Annarita D'Addabbo [1] and Marco Chini [3]

[1] National Research Council—Institute for Electromagnetic Sensing of the Environment (CNR-IREA), 70126 Bari, Italy; annarita.daddabbo@cnr.it
[2] Earth and Geoenvironmental Science Department, University of Bari, 70125 Bari, Italy; marina.zingaro@uniba.it
[3] Environmental Research and Innovation Department (ERIN), Luxembourg Institute of Science and Technology (LIST), L-4422 Belvaux, Luxembourg; marco.chini@list.lu
* Correspondence: alberto.refice@cnr.it

Received: 4 September 2020; Accepted: 28 September 2020; Published: 30 September 2020

Abstract: Flood detection and monitoring is increasingly important, especially on remote areas such as African tropical river basins, where ground investigations are difficult. We present an experiment aimed at integrating multi-temporal and multi-source data from the Sentinel-1 and ALOS 2 synthetic aperture radar (SAR) sensors, operating in C band, VV polarization, and L band, HH and HV polarizations, respectively. Information from the globally available CORINE land cover dataset, derived over Africa from the Proba V satellite, and available publicly at the resolution of 100 m, is also exploited. Integrated multi-frequency, multi-temporal, and multi-polarizations analysis allows highlighting different drying dynamics for floodwater over various land cover classes, such as herbaceous vegetation, wetlands, and forests. They also enable detection of different scattering mechanisms, such as double bounce interaction of vegetation stems and trunks with underlying floodwater, giving precious information about the distribution of flooded areas among the different ground cover types present on the site. The approach is validated through visual analysis from Google Earth™ imagery. This kind of integrated analysis, exploiting multi-source remote sensing to partially make up for the unavailability of reliable ground truth, is expected to assume increasing importance as constellations of satellites, observing the Earth in different electromagnetic radiation bands, will be available.

Keywords: flood monitoring; ALOS 2; Sentinel-1; multi-sensor integration; multi-temporal inundation analysis; Zambesi-Shire river basin

1. Introduction

Satellite remote sensing plays an important role in the observation of flood events [1–3]. Synthetic aperture radar (SAR) imagery is particularly useful for water extent detection [4–6], thanks to its all-weather, day/night imaging capabilities. The availability of frequent SAR acquisitions is enabling unprecedented timeliness and accuracy in modeling and monitoring of inundation phenomena [7–10]. A further advantage of SAR sensors is the possibility of better recognizing floodwater in different ground conditions, thanks to their insensitivity to confusing factors such as water color, and the high sensitivity of the microwave radiation to water surfaces. The latter determines the appearance of open, calm water as dark in a SAR image; moreover, SAR often permits detecting water beneath vegetation, thanks to the capacity of microwaves to penetrate below the vegetation canopy. This allows detecting the double-bounce mechanism that increases with the presence of water under vegetation [11]. Various parameters affect this scattering mechanism depending on radiation features (wavelength and polarization) and surface conditions (vegetation height, incidence angle, water level, and soil moisture) [12]. For instance, penetration under foliage of vegetated canopies increases with wavelength, so that L-band sensors (wavelength of about 24 cm) are more sensitive than C-band sensors (wavelength of the order of 5 cm) [13]. The latter characteristic renders L-band data more and more attractive for the monitoring of wetlands [14,15].

The investigation of inundation phenomena through SAR data on vegetated areas is often performed through an integrated analysis [16,17]: different spectral bands can be exploited to identify distinct backscattering mechanisms. The combination of analytical techniques and data overlap can help determine the response of flooded areas with distinct vegetation cover to the microwave signal. This is useful especially in cases, which actually constitute the majority, in which ground data are scarce or not available. In fact, availability of ground truth during inundation events is a rare occurrence, mainly due to the typical short warning times, and the difficult situation on the ground caused by the meteorological conditions and the flood events themselves, especially in less developed countries, where access to particular areas can be even more difficult. In such cases, integration of several data sources, heuristic inference, and data processing techniques can often make up for missing ground truth, allowing to retrieve significant information about the types of land cover and how they are affected by the flood [18].

The present study investigates the application of multi-temporal, multi-frequency, and multi-polarization SAR data, in synergy with globally-available land cover data, for improving flood mapping in vegetated areas. The Zambezi-Shire area features a variegated surface cover: wetlands, open and closed forest, cropland, grassland (herbaceous and shrubs), and a few urban areas. The presence of low and high vegetation (typical of the tropical landscape), and the alternated proximity of bare soil or scarcely vegetated areas requires interpreting the behavior of different land cover classes in different conditions (flooded/not flooded). We show how the combination of various analytical techniques and the simultaneous availability of data with different frequencies and polarizations can help to recognize the response of flooded areas with distinct vegetation cover to the microwave signal. This integrated approach is finalized to explore and refine information increase and data synergy for flood mapping.

We focus on a particular event occurred in late January–early February 2015 in an area located between the Zambezia and Tete provinces, Mozambique. We select SAR images acquired in L and C band (ALOS 2 and Sentinel-1, respectively), before and during the event, in order to analyze the spatial and temporal evolution of the inundation by taking advantage of their different wavelength and polarization. We show how C-band images can be integrated with dual-polarized, L-band ALOS 2 images, co- and pre-event; this helps highlighting different responses of flooded areas to the radar signal, caused by diversified land cover and synergy between different wavelengths and polarizations when acquired simultaneously. L-band images have a higher penetration within the forest canopy than C-band images, and, in most cases, cross-polarized (HV or VH) signals have a lower double-bounce effect than co-polarized ones (HH and VV) on vegetated areas, where e.g. HV stands for horizontally-polarized transmit, vertically-polarized receive backscattering, etc. [19]. This diversity (in wavelength and polarization) gives more information about the scattering mechanism of the surface, and therefore contributes to isolate different scattering classes thus better recognizing land cover types [3,12].

In our study, we analyze phenomena of backscattering decrease and increase, determining flooded and non-flooded areas through heuristic interpretation of different backscattering behaviors, isolated through K-means clustering and compared to classes from the CORINE land cover database. Results are analyzed in correspondence with optical Google Earth™ imagery and red-green-blue (RGB) channel combinations of SAR data. In this way, the recognition of flooded regions is combined with the information of the type of vegetation, supporting the interpretation of interaction effects between the water surface and the radar signal.

By considering this starting point (i.e., the potential of diverse SAR data) and the characteristics of the region (i.e., the confluence of Zambezi-Shire rivers, recurrently flooded, characterized by short and tall vegetation), this study represents a test case to experiment the value of multi-source SAR data integrated with medium resolution land cover data for distinguishing flooded areas on various land cover types, without availability of full-fledged ground truth data acquired on the field.

2. Study Area

The Zambezi river basin, one of the largest in Africa, has a complex morphology and a heterogeneous surface cover. These characteristics and the repeated occurrence of flood events in the catchment make it particularly suitable for the monitoring of flooded areas through multi-temporal analysis. Several studies investigated the interrelation between the hydro-morphologic features of the Zambezi River, their evolution over time and the periodic inundations, by exploiting Earth observation data [20–23].

The basin, large about 1.4 million km^2, is located in south-eastern Africa (Figure 1a) and spreads over eight countries (Angola, Botswana, Malawi, Mozambique, Namibia, Tanzania, Zambia, and Zimbabwe). The fluvial system is composed by the Zambezi River with its major tributaries (Congo, Cuando, Kafue, Luangwa, and Shire) and by the two dams of Kariba and Cahora Bassa. The course of the river can be divided into three main segments that represent different geomorphological units—Upper, Middle and Lower Zambezi—in which the river morphology changes in relation to the physiographic regions. Between the towns of Mutarara and Chimuara, the Lower Zambezi is characterized by the confluence with the main tributary, the Shire River, whose headwaters are in Lake Malawi. From the lake, the Shire River flows southwards, traversing gauges, rapids, and waterfalls until it forms a broad floodplain extending from Chikwawa to the confluence with the Zambezi, crossing the Elephant and Ndindi marshes and the Ilha de Inhangoma region. The latter, created by splitting the Zui Zui channel into the Shire River after a harmful flood in 1840, represents a region of interest, because recurrently subject to floods (Figure 1b). In fact, when the Zambezi is in flood, the (channeled) overflow pours into the Shire increasing its streamflow, until it exceeds and floods the valley [24]. This condition occurs with the reaching of peak flow (January–March) in the Zambezi and Shire rivers, during the wet season (November–April). The hydrology of the fluvial system, in fact, is determined by the seasonality of rainfall and water levels patterns, which, however, are conditioned by climate change and anthropic impact (flow regulation, i.e., dams) [23,25,26]. The variability of the natural processes (morphologic, hydrologic, and climate dynamics) has been modified strongly during the last decades, influencing the predictability of flood cycles and the consequent natural environment and anthropic landscape (villages, croplands, human activities, etc.) [27,28]. Moreover, as mentioned in the introduction, a variegated surface cover characterizes the area.

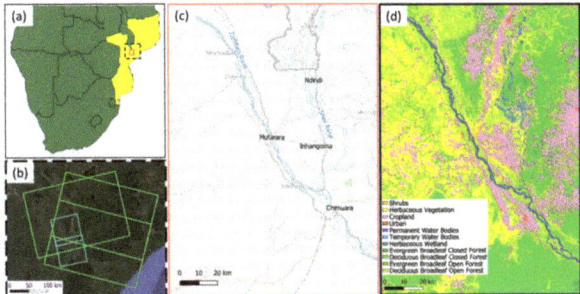

Figure 1. (**a**) Location of the Zambezi River basin, (**b**) enlarged view with footprints of the Satellite imagery—green: Sentinel-1 (ascending and descending), cyan: ALOS 2, (2 frames); (**c**) detail map with locations and toponyms described in the text; (**d**) CORINE land cover map for the region of interest, with legend.

3. Materials

ALOS-2 PALSAR 2 (A2) and Sentinel-1 (S1) data were collected over adjacent, partially superposed orbits, in the dates and with the acquisition parameters listed in Table 1. Acquisition footprint locations are shown in Figure 1b. The two A2 scenes (two frames per date) are acquired in L band, in the FBD (Fine-Beam Double polarization) mode, in the HH and HV polarization channels, with ~10 × 10 m^2 ground resolution, in ascending geometry. S1 images are acquired in C band, in the IW (Interferometric Wide-Swath) mode, in VV polarization, with 20 × 5 m^2 (azimuth × range) resolution, in both ascending and descending geometries. Incidence angles are roughly comparable for all the used imagery, with 32° for A2 and 36° to 42° for S1 (S1 IW images have more than 15 degrees variation in incidence angle from ~30° at near to ~46° at far range: here we approximate the values for the selected area within the wide swath frame).

As can be seen from Table 1, one A2 acquisition is available on 9 February 2015, during the event, while one pre-flood image was acquired on 1 December 2014. S1 data are acquired with a 12-days repeat cycle in the period of interest. In our case, both ascending and descending acquisitions are considered, with intervals of 12 and 6 days. The higher frequency of acquisition of S1 images allows to better follow the event, with one acquisition in late January and 4 acquisitions in February 2015. One additional acquisition on 22 April 2015, is used as reference. All intensity images were calibrated, speckle filtered [29], geocoded and converted to dB scale.

We also use the CORINE land cover database [30], available at a resolution of 100 × 100 m^2 over the African continent to interpret the backscattering evolution in time at different frequencies and polarizations. Although obtained by temporally averaged data from the Proba V satellite, such map contains sufficient information to characterize terrain typologies. The portion of the CORINE land cover map for the study area is shown in Figure 1c, with its standard color map legend.

All data were resampled to a ground resolution of 20 × 20 m^2 through a nearest neighbor algorithm.

Table 1. Details on the used imagery. A = ascending, D = descending geometry. Images indicated in red bold text color are those selected for the subsequent multi-frequency analysis.

Sensor	Date	Polarization	Geometry
ALOS 2	**1 December 2014**	**HH, HV**	**A**
Sentinel-1	29 December 2014	VV	D
Sentinel-1	22 January 2015	VV	D
Sentinel-1	**3 February 2015**	**VV**	**D**
ALOS 2	**9 February 2015**	**HH, HV**	**A**
Sentinel-1	15 February 2015	VV	D
Sentinel-1	21 February 2015	VV	A
Sentinel-1	27 February 2015	VV	D
Sentinel-1	22 April 2015	VV	A

4. Methods

4.1. Preliminary Analysis—Detection of Open Floodwater as Decrease of Backscattering Value

One important aspect of flood monitoring is the possibility to follow an event in time. This is achievable at basin or larger scales through use of remotely sensed data acquired with high temporal frequency. SAR data are particularly sensitive to drops in microwave backscattering due to the presence of open water on the terrain surface. SAR data time series can be thus exploited, even in synergy with e.g., optical data [31], to compute multi-temporal maps illustrating the evolution of an event, such as the progressive draining of flooded areas, according to the topography and other hydraulic terrain characteristics [8,18].

Determining the extent of open water in a multi-temporal stack can be difficult when dealing with vegetated areas, as absolute backscatter drops may depend, e.g., on local conditions of wind or vegetation height. Several solutions have been proposed to retrieve reliably open water in time series of SAR images, relying, e.g., on harmonic analysis to model periodic flooding [7], or interpretation of backscatter time signatures [9]. We use clustering, a well-known methodology for data exploration and analysis. We adopt one of the best-known algorithms for data clustering, i.e., K-means. The K-means algorithm [32] iteratively assigns each element of a multi-dimensional feature space (the pixels' backscatter values in our case), assumed Gaussian-distributed, to one among a pre-defined number of clusters, choosing the one with the shorter (Euclidean) distance, then recalculating cluster centers until convergence.

Here, we perform clustering of pixel backscatter values on the multi-temporal stack of images. We use a relatively large number of clusters (K = 32), to avoid neglecting small clusters of pixels with distinct behavior. We then inspect the retrieved clusters spatially and spectrally (i.e., the backscatter values of the cluster centers), to identify areas undergoing flood in each image. The clustering helps focusing on pixel sets which better follow expected backscatter change with respect to an unflooded image, neglecting other uninteresting typologies. This approach partly follows the core methodology delineated in the DAFNE algorithm [33], with a few differences. First of all, the automatic computation of prior probability values for the various clusters when L band signals are considered is not yet implemented in the DAFNE toolbox, so a heuristic procedure is used for this step. In fact, the experimental results here obtained can be considered as a useful study to opportunely integrate the software. Then, we do not have reliable ancillary data to precisely constrain spatially the flood phenomenon around the river course, as wetlands and other flooded terrain types are spread over a rather large surface in the river basin.

We here select the clusters with centroid values which best represent expected backscatter levels for each typology of flooded/non-flooded terrain, thus obtaining a multi-temporal representation of the flood evolution by considering areas inundated at different dates. Such multi-temporal flood maps are finally "classified" by assigning them to their corresponding CORINE class, in order to understand how the open floodwater is distributed over the different land cover classes.

The left panel in Figure 2 shows the multi-temporal flood map obtained from S1 data. The map shows several areas around the main Shire river course, which are flooded only on the first imaged date of the event, i.e., 22 January 2015 (in light blue), and then gradually dry, so that smaller areas appear flooded from the first to each of the subsequent dates, from 3 February 2015 until 27 February 2015 (in darker blue tones). The darkest blue areas are those identified as water in the flood-free image acquired after the event, on 22 April 2015, taken as reference. Note that the Shire river course, clearly visible in the bottom part of the map, appears correctly in dark blue.

The right panels show the six successive flood maps derived from the S1 acquisitions, colored according to the underlying CORINE land cover classes, reported in the legend in Figure 1. It can be seen that the largest flooded area, in the first date, covers several class types, spanning cropland (pink pixels, spread throughout the whole region), herbaceous wetlands (blue-green areas at the center/bottom), all four types of broadleaf forest (evergreen/deciduous, open/closed, corresponding to different shades of green, present especially close to the center-right of the area), and herbaceous vegetation (yellow areas mostly in the bottom-left). As the flooded area shrinks, in the following dates, it affects less and less herbaceous vegetation and cropland, leaving mostly wetlands and a few forested areas as flooded in the last date.

This is confirmed by analyzing quantitatively the number of pixels detected as (open-water) flood in each CORINE class, for each of the S1 acquisitions, as reported in Figure 3 It can be seen how, on the first event date, the CORINE class most affected by flood is the Deciduous Broadleaf Open Forest, followed by Wetlands, Cropland, and Herbaceous Vegetation. The quantity of flooded pixels in the Deciduous Broadleaf Open Forest class decreases, during the evolution of the event, to approximately 1/5 at the end of the period. Herbaceous Wetland pixels are the second most populated class in the first date, but they become the first one from the second date onward. Cropland, Herbaceous Vegetation and Deciduous Broadleaf Open Forest all show similar decreasing trends, while Deciduous Broadleaf Close Forest shows smaller areas throughout the whole imaged period. Other classes appear more marginally affected. Notably, the areal extent of both the Permanent and Temporary Water Bodies classes remain practically constant throughout the event, thus qualitatively confirming the consistency of the analysis.

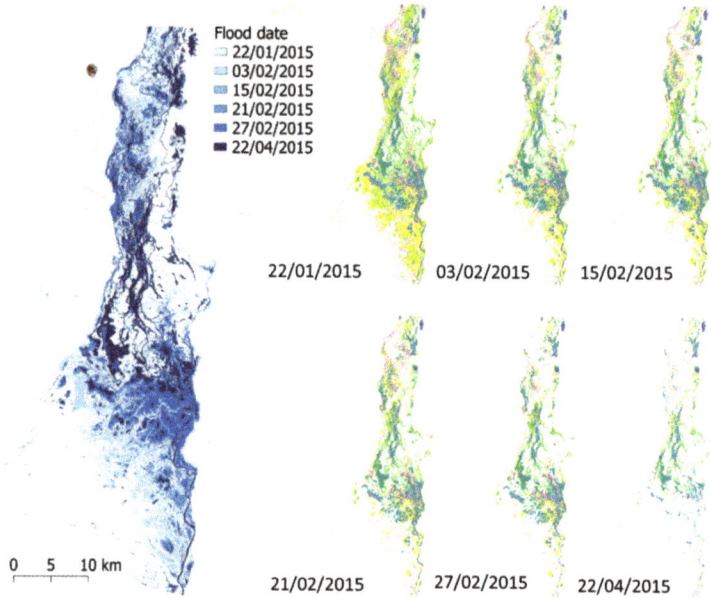

Figure 2. Left: multi-temporal flood map for an area of the Shire basin, obtained by Sentinel-1 images. Areas with darker colors are flooded for longer periods, thus depicting the shrinking of the flooded surface from the first (22 January 2015), up to the last flood date (27 February 2015) reported in the legend. Darkest areas are covered by permanent water (at the post-event date of 22 April 2015). Right: maps of flooded areas on each of the 6 S1 acquisition dates, with colors corresponding to the CORINE land cover classes (shown in Figure 1).

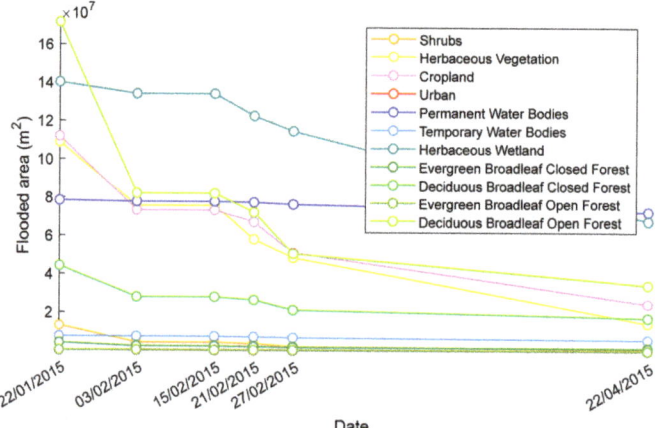

Figure 3. Estimation of the areas flooded for each CORINE land cover class, in the sequence of Sentinel-1 images. Colors in the legend follow the standard CORINE class labels shown in Figure 1.

It can be noticed that many areas at the center-right of the maps in Figure 2 appear not flooded in any of the acquisition dates, thus looking like "holes" in the maps. By looking at the overall CORINE map in Figure 1b, these appear to correspond mostly to forested areas. This observation is thus compatible with the interpretation that C-band radiation is not able to penetrate thick ("closed", in the CORINE terminology) forest cover. However, many of the closed forest undetected areas appear to be spatially close, or even surrounded, by areas detected as flooded. As no significant topography is present in the area, it is presumable that at least part of these forested areas are indeed flooded beneath the canopy, although C-band microwaves cannot "see" through the thick forest stands.

One last observation about the C-band multi-temporal analysis is that, as appears in both the maps in Figure 2 and the plots in Figure 3, the flooded area between 3 and 15 February 2015 appears identical. In our K-means clustering analysis, this comes from the fact that it was not possible to determine statistically significant pixel clusters showing different flood behavior on these two dates. In fact, an RGB combination of the two Sentinel-1 intensity images acquired on 3 February 2015 (red channel) and 15 February 2015 (green and blue channel), reported in Figure 6a below, shows that very little appears to change from one date to the other, as very few red or cyan colored pixels are visible within the area interested by the flood. This allows to confirm empirically that, between these two dates, the situation on the ground is substantially stationary, as suggested by the previous multi-temporal analysis. As a consequence, the available A2 image, acquired on 9 February 2015, can be investigated in synergy with one of the two S1 acquisitions, treating the S1-A2 imagery as a multi-frequency dataset referred essentially to the same snapshot in time. This is investigated in Section 4.2 below.

The same analysis as the one just presented on C-band data has been performed for the L-band stack, composed by only two images, one acquired before and one during the flood event. In this case, analysis is limited to the "permanent water" areas detected in the pre-event image, and the open water areas in the single co-event image. Figure 4 shows on the left the multi-temporal map in the same color code as the one on the left of Figure 2, and on the right the two maps corresponding to the two dates, with each pixel classified and colored according to the corresponding CORINE map. By comparing the multi-temporal maps in Figures 2 and 4, some differences can be noticed. First of all, as expected, the multi-temporal map obtained from A2 data appears slightly more "dense" (i.e., with less empty areas) in the central part of the basin with respect to that obtained from S1.

Another observation concerns the distribution of land cover classes affected by the flood in the two cases. As noted above, S1 data allow to conclude that the situation on the ground between the dates of 3 and 15 February is substantially the same; nevertheless, the flood map realized from A2 data, dated 9 February (the rightmost map in Figure 4), appears to have different quantities of yellow (corresponding to herbaceous vegetation), pink (corresponding to crops) and light green (deciduous open forest) pixels than the corresponding ones from S1. This is confirmed by looking at the estimated areas covered by open flood water for each CORINE class in the two A2 dates, reported in Figure 5. A higher area covered by open forest can be noticed, and at the same time smaller areas corresponding to herbaceous vegetation and cropland. This comparison, taking for valid the assumption of stationary ground situation between the two dates of 3 and 15 February, shows that the interaction and backscattering of L-band microwaves with flooded terrain is slightly but significantly different than that pertaining to C-band radiation.

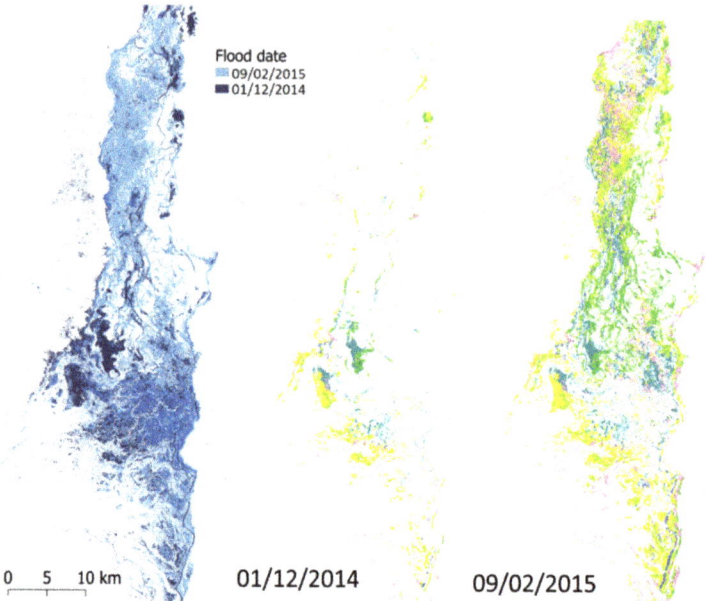

Figure 4. **Left**: multi-temporal open-water flood map for the Shire basin, obtained by ALOS-2 images. Areas with darker colors are flooded for longer periods, thus depicting the evolution of the flooded surface from the first, pre-event to the second, co-event image, as reported in the legend. **Right**: maps of flooded areas on each of the 2 A2 acquisition dates, with colors corresponding to the CORINE land cover classes (shown in Figure 1).

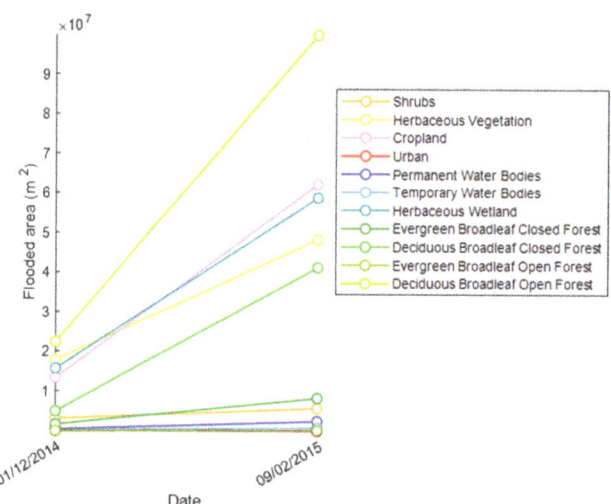

Figure 5. Estimation of the areas flooded for each CORINE land cover class, in the sequence of Sentinel-1 images. Colors in the legend follow the standard CORINE class labels.

To further delve in this matter, in Figure 6b we show a RGB combination of the SAR backscatter of the A2 (HH) images acquired on 1 December 2014 (red channel), and on 9 February 2015 (green and blue channels): as can be seen, some groups of red pixels, which are the areas exhibiting lowering of backscatter in L band, correspond to the ones which are dark in Figure 6a, i.e., open flood water in both the co-event C-band images. However, some cyan areas are also visible in correspondence of C-band dark areas. Moreover, some areas in the lower-left part of the basin appear bright in C band (panel a) and dark in L band (panel b). These large differences between the two images confirm the presence of areas with different penetration and thus different microwave interaction in images from the two sensors.

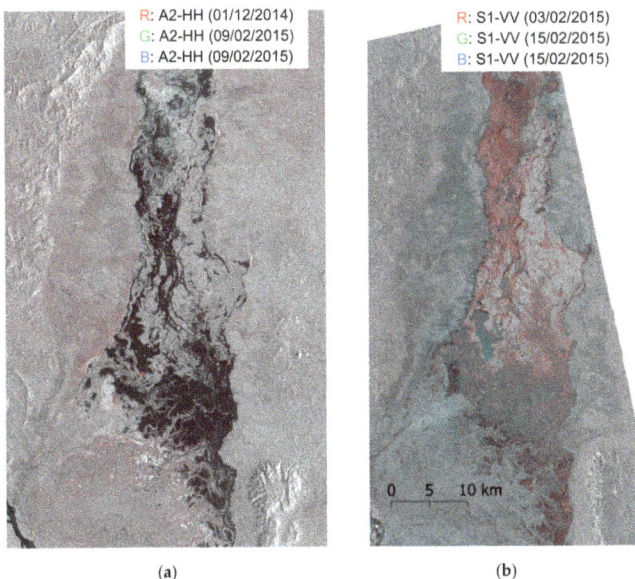

Figure 6. (a) RGB combination of the two S1 images acquired on 3 (red) and 15 (green and blue channels) February 2015. (b) RGB combination of the A2 images acquired on 1 December 2014 (red) and 9 February 2015 (green and blue channels).

4.2. Statistical Multi-Sensor Backscatter Analysis of Land Cover Classes

We now adopt the dataset shown in red color in Table 1, which includes one pre-event and one co-event image for each type, one S1 pair in VV polarization, and two A2 pairs in HH and HV polarizations, respectively, as a multi-sensor stack referring to the same situation on the ground. To gain some further insight into the different types of land cover present on the ground, we analyze statistically the backscatter signatures of the various CORINE land cover classes. In Figure 7 we report the histograms of the SAR intensities for pixels belonging to the land cover classes present over the area, selected by using masks obtained by the CORINE map.

As can be seen, Shrubs and Cropland exhibit a slight increase of backscatter (the peaks move to higher values) from pre- to co-event imagery in C band (up to 1 dB), but a more consistent one in L band (3–5 dB). This could be due to the presence of double bounce effects in flood conditions, which is known to increase backscatter up to several dB. A moderate increase of backscatter (peaks shift of about 1 dB in both bands) is also detected on the Urban class, although sample population is much reduced in this case. Deciduous Broadleaf Open and Closed Forest classes exhibit a less uniform behavior, with an appreciable increase registered only in L band and HH polarization (A2-HH), while in both L-band HV and C-band VV images, the intensity distributions do not change consistently. Open forest seems to exhibit smaller secondary peaks at very low backscatter levels (approximately between −20 and −25 dB), possibly corresponding to flooded forest patches. Similar lower secondary peaks are just barely visible in pixels corresponding to closed forest, as could be expected from a lower penetration of the thicker canopy.

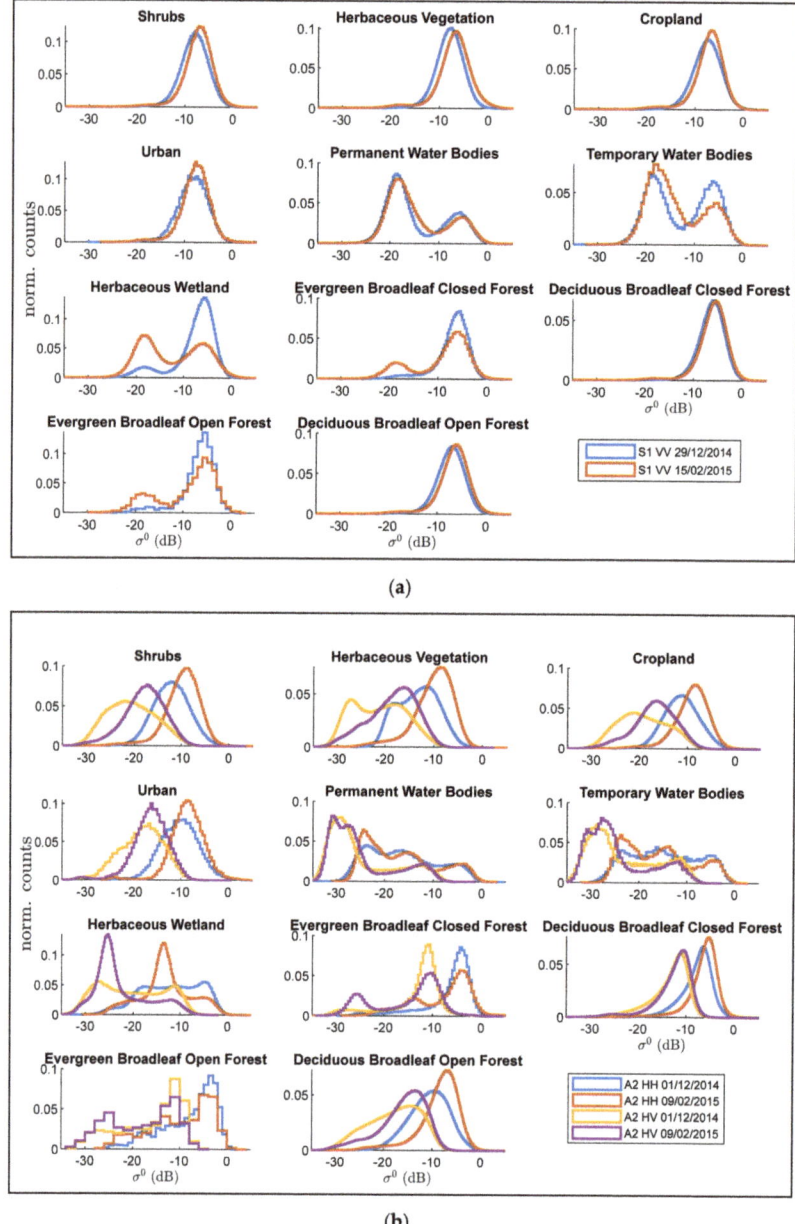

Figure 7. Histograms of synthetic aperture radar (SAR) intensities over the CORINE land cover classes found on the image, for (**a**) C band, (**b**) L band.

Some types of land cover exhibit clearly bimodal intensity distributions in at least some of the SAR image bands. This is evident for classes such as Herbaceous Vegetation (in L band), and Herbaceous

Wetland (in both bands). The former exhibits just a small (~1 dB) increase in the peak backscatter value in C band; in L band, pre-flood backscatter exhibits a peak at very low values in both polarizations, as well as a second one at higher values, while in co-event images the first peak appears much lower, and the secondary peak shifts up by 2–3 dB. The latter (Herbaceous Wetland) exhibits a clear increase in the low-intensity peak in C band from pre- to co-event data; in L band, both polarizations show a consistent increase and a small shift of the low-intensity peak from pre- to co-event. Both Open and Closed Evergreen Broadleaf Forest show bimodal histograms in co-event images, while pre-event data show very small to null low-intensity peaks.

Bi- or sometimes tri-modality can also be observed for Permanent and Temporary Water Bodies, with small shifts from pre- to co-event conditions. The presence of several peaks is likely due to the lower resolution of the CORINE data with respect to the SAR ones, so that many pixels falling in the Permanent/Temporary Water Bodies CORINE class (mostly located on and around the river course) actually correspond to other land cover classes on the ground; the lower peak in C-band VV backscatter seems to correspond to two separate peaks in L-band HH data, while L-band HV is still bimodal, although with distributions shifted by −6/−10 dB.

These results already confirm a consistent increase in information brought about by the use of multi-frequency data. As is often the case, especially for remote regions such as those in African countries, where validation or updating campaigns are difficult, the land cover map may contain some difference with respect to the actual situation on the ground. To further refine our inference, we analyze in higher detail the S1-A2 integrated multi-frequency dataset.

4.3. Multi-Frequency and Multi-Polarization Floodwater Mapping

As mentioned above, the backscattering properties of terrain surfaces, as well as their changes due to inundation events, depend on the wavelength of the radar sensor. In thickly vegetated areas, floods may affect different types of ground environments, exhibiting different vegetation densities and other terrain properties, so that detecting exhaustively floodwater in such cases can be difficult. For instance, it is important to be able to detect flooded vegetation, besides open water areas. This is possible thanks to the so-called double bounce phenomenon, in which microwaves can be scattered back towards the sensors after bouncing on the surface and the vertical vegetation structures (tree trunks or plant stems). In other cases, more sparse and thin vegetation, such as grass or shrubs, may appear darker when flooded in SAR images acquired at longer wavelengths, as the diffusion effect of the canopy is weaker. Vice versa, shorter wavelengths may give rise to enhanced backscatter from such vegetation in flood conditions.

After the preliminary analyses described in the previous sections, we are now ready to use the complementarity of the A2 L-band image acquired during the flood, with both its polarization channels (HH-HV), and the S1 C-band one, to gain some further insight into the different types of land cover present on the ground and their backscattering response in presence of floodwaters, thanks to the multi-frequency (and multi-polarization) stack of C- (VV) and L-band (HH and HV) pre- and co-flood image built in the previous sections. This stack presents a variety of behaviors, including backscatter decrease and increase for various vegetation types and flood conditions, caused by phenomena such as specular reflection, double bounce, or wind. To better interpret such huge wealth of information, this stack is again processed through K-means with a number of clusters sufficient to isolate backscatter change signatures. The number of clusters is then reduced through a merging procedure which iteratively joins the two clusters with minimum Battachaarya distance. Finally, the most interesting clusters are interpreted by comparing their signatures and change with the CORINE land cover information. The algorithm is thus composed by the following steps:

1. Full data stack (multi-temporal, -frequency, and -polarization) clustering, where the first guess for clusters number is the double of land cover classes available. In our case, the number of available classes is provided by CORINE land cover map and each class is assumed to be flooded

or non-flooded. The only exception is the permanent water class that cannot change its state during the flood event.
2. Clusters number reduction.
3. Classification of each cluster based on the mean values of input features and the dominant land cover class.

5. Results

As described above, our data are arranged in a 6-dimensional array, containing the six SAR intensities corresponding to the entries in red in Table 1. We begin our analysis by heuristically choosing a number of clusters equal to 21, which corresponds to the number of CORINE classes present on the area (11), considered in both flooded and unflooded conditions, with the obvious exception of permanent water areas.

Figure 8 shows a map of the clustered image, spanning the whole area covered by the image frames, where pixels clusters are represented in different colors. To ease interpretation, 3 scatter plots are also shown in the figure, showing the 21 cluster center positions in each of the 3 planes spanned by the pre-flood and the flood backscatter values in each channel (indicated as times t_0 and t_1, respectively). As can be seen, the map shows some cluster centers which appear relatively compact spatially, while others are more scattered throughout the area. In the scatter plots, cluster centers which fall farther from the main diagonal (shown as a dashed line in each plot) exhibit the most interesting behaviors.

Some cluster centers appear above the scatter plots' diagonals, at least in some cases. These are likely to correspond to double bounce phenomena due to flooded vegetation, leading to increased backscatter in one or more channels. It appears, however, that many cluster centers of this kind are very close to each other, suggesting their possible belonging to the same class on the ground. In practice, it appears that the postulated initial number of clusters (21) is too high with respect to what can be actually distinguished in the data, and thus a lower number may be more acceptable. However, which one is the best? The choice of the most suitable cluster number is a long-standing problem in data analysis, and several methods have been developed to deal with it. A common practice is to start from a relatively large number of clusters, then merge "close" clusters iteratively, according to some distance measure, until satisfying some quality criterion. Such a hierarchical approach has been adopted, e.g., in [34], by using available ground truth as a reference. Here, we adopt a similar approach, stopping the clustering procedure when a heuristic criterion of homogeneity is reached.

We proceed as follows: starting from the maximum number of clusters (21 in this case), we merge together the two clusters which have the smallest Bhattacharyya distance [35] from each other, iterating the process until being left with only two clusters encompassing the whole dataset. At each iteration, we re-compute the coordinates of the cluster centers (as average of the coordinates of the points belonging to that cluster).

In Figure 9 we show the minimum Bhattacharyya distance computed between the various clusters, as a function of the number of clusters. The measure stays relatively constant as clusters are decreased from 21 to about 12 (moving from right to left in the plot), then has a general increasing trend as clusters are reduced further, with some fluctuations, reaching the maximum value for just 2 clusters. We take the position of the approximate discontinuity in the trend described above as the "optimal" value, through the following reasoning: while pairing clusters above the "optimal" number, the minimum distance is only slightly affected—i.e., we are likely merging bulks of very close clusters. As the last cluster of these bulks is merged with its closest member, then larger distances begin to be left in the data, and thus the minimum Bhattacharyya distance increases. So, our "optimal" cluster number is the one after which the minimum distance begins to increase appreciably. This corresponds, in our case, to the number of 12 clusters.

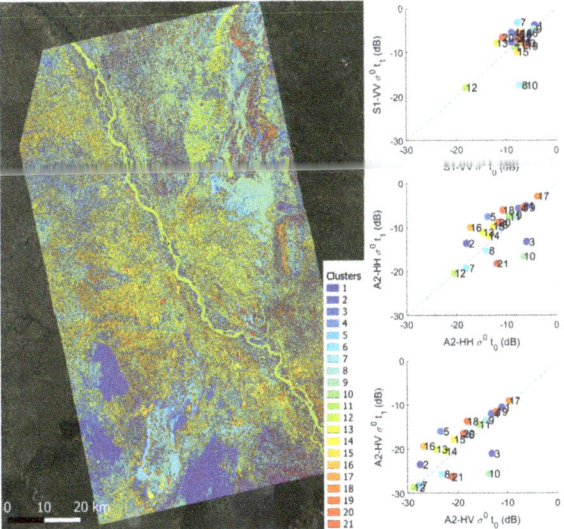

Figure 8. Left: K-means cluster map obtained with K = 21 clusters. **Right**: the three scatterplots show the cluster centers in the 3 planes whose axes are the 3 channels backscatter values in the pre- and flood acquisition dates. Cluster numbers are reported next to each colored dot to facilitate recognition in the legend and thus in the map.

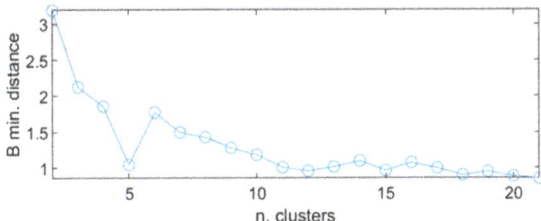

Figure 9. Minimum Bhattacharyya distance as a function of the number of clusters in the multi-frequency dataset.

Figure 10 then shows the corresponding clustered map, together with the same scatter plots as in Figure 8, for this "optimal" number of 12 clusters. Here, more uniformly spaced cluster centers are visible. In addition to those corresponding to permanent water (n.9 in this case), and open water flooded areas (n.8), we notice cluster n. 11, which corresponds to an increase of about 8 dB in L band, while exhibiting no significant change in C band, so likely corresponding to forested areas with thick canopy layer, which can be penetrated by lower frequency electromagnetic waves, thus causing double bounce with the bottom water layer, but not by C-band radiation.

This cluster is represented in bright red and appears in the top part of the map on the left of Figure 10, corresponding loosely to forest classes in the CORINE database. A similar behavior, although with lower intensity increases (up to a maximum of 3–4 dB) can be discerned for clusters n. 2, 4, and 10. A rather peculiar behavior is shown by cluster n. 6, in cyan-greenish color, which exhibits an increase of about 4–5 dB in C band, while staying very close to the diagonal, with rather low backscatter values at both times, in L band. This class, located in a rather compact area at the center of the map, likely corresponds to shrubs or low, thin vegetation, causing likely an increase in C-band backscatter, while resulting "transparent" to longer wavelength radiation.

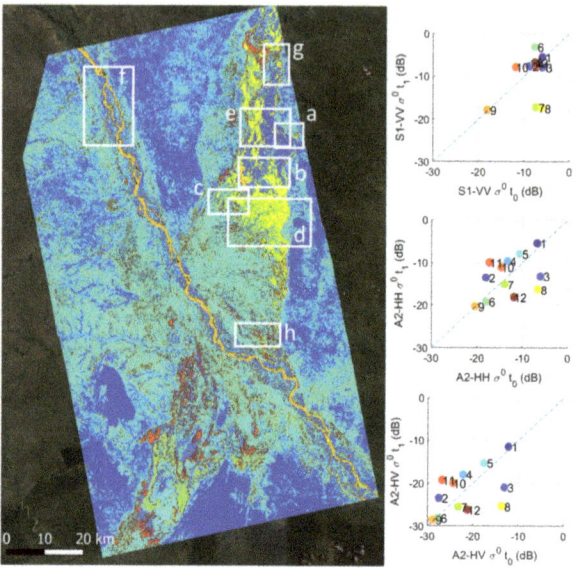

Figure 10. As in Figure 8, with K = 12 clusters. White rectangles on the left map, labeled from a to h, highlight detail areas shown enlarged in the subsequent figures.

Figure 11 illustrates the correspondence of the 12 clusters with the classes in the CORINE land cover map. Each entry in the matrix represents the percentage of pixels belonging to a given cluster (columns) and a given CORINE class (rows). This matrix offers some additional indication about the clustering results. In the following, we present an integrated interpretation of the types of response of each cluster with respect to the corresponding land cover CORINE classes.

	1	2	3	4	5	6	7	8	9	10	11	12
Shrubs	1.084	4.232	1.706	3.899	2.736	1.934	1.624	0.8643	1.082	5.303	3.531	2.836
Herbaceous Vegetation	6.74	44.39	6.53	20.81	16.6	53.39	21.96	11.67	3.913	25.22	40.84	16.23
Cropland	13.24	20.17	17.05	31.63	24.17	5.584	19.63	18.33	3.996	33.95	20.1	21.69
Urban	0.09053	0.06615	0.02189	0.1564	0.1916	0.007893	0.06547	0.01527	0.1011	0.1638	0.08074	0.01547
Permanent Water Bodies	0.6186	0.5269	2.463	0.3599	0.3426	0.5571	3.014	1.688	47.99	0.4511	0.1223	2.177
Temporary Water Bodies	0.07331	0.0893	0.3858	0.04719	0.04255	0.05656	0.7538	0.6383	4.003	0.05974	0.02493	0.3179
Herbaceous Wetland	0.9526	3.89	11.6	0.9165	0.4663	4.866	24.05	22.87	12.5	0.6271	1.958	5.985
Evergreen Broadleaf Closed Forest	0.6878	0.07524	2.347	0.05749	0.08283	0.04034	1.13	2.702	0.8538	0.03747	0.04588	0.1109
Deciduous Broadleaf Closed Forest	40.44	1.61	17.79	5.76	12.84	2.063	5.622	14.95	4.603	3.169	2.453	6.515
Evergreen Broadleaf Open Forest	0.01691	0.01251	0.05863	0.005684	0.004059	0.008331	0.04075	0.09552	0.03733	0.003387	0.008839	0.02262
Deciduous Broadleaf Open Forest	36.06	25.14	40.00	36.36	40.53	31.89	22.08	26.19	20.92	31	30.84	44.1

K-means class

Figure 11. Class pixel relative populations for the multi-temporal K-means classified image.

To improve the visual representation of local spatial support of the backscatter signatures, we also consider a RGB combination of the three ratios between co- and pre-event SAR, red for the S1-VV, green for the A2-HH, and blue for the A2-HV (Figure 12). The different speckle patterns in the three channels give this image a somewhat smoother appearance than single SAR images, so we use the color in this image as an aid to detect and interpret the multifrequency type of backscatter with respect to the land cover. Generally, in the change RGB image in Figure 12, black areas underwent strong backscatter decrease (colors are saturated at ±10 dB for better contrast visualization, as shown in the figure inset color cube) in all three channels, likely corresponding to open water; dark red corresponds to decreasing backscatter in L-HH/HV imagery, while maintaining roughly constant levels in C-VV. This would likely correspond to vegetation with a structure which allows penetration of longer wavelengths (L band), which therefore undergoes specular reflection, while C-band waves are backscattered by the canopy, and therefore do not exhibit significant changes when flooded. Dark green pixels underwent backscatter decrease in L-HV and C-VV, while keeping roughly constant values in L-HH. This could be due to different wind conditions on the two acquisition dates (3 and 9 February 2015 for A2 and S1, respectively), which cause the water surface to backscatter more power in the second date (in L-band) than in the first one (in C-band). The higher change in HH than in HV polarization also suggests this kind of explanation, since cross-polarized channels are reported to be less sensitive, in terms of backscatter, to rough surfaces such as water interested by capillary waves. Notably, the opposite behavior (decreasing HH, constant HV channels), which would correspond to dark blue pixels, is not seen on the image.

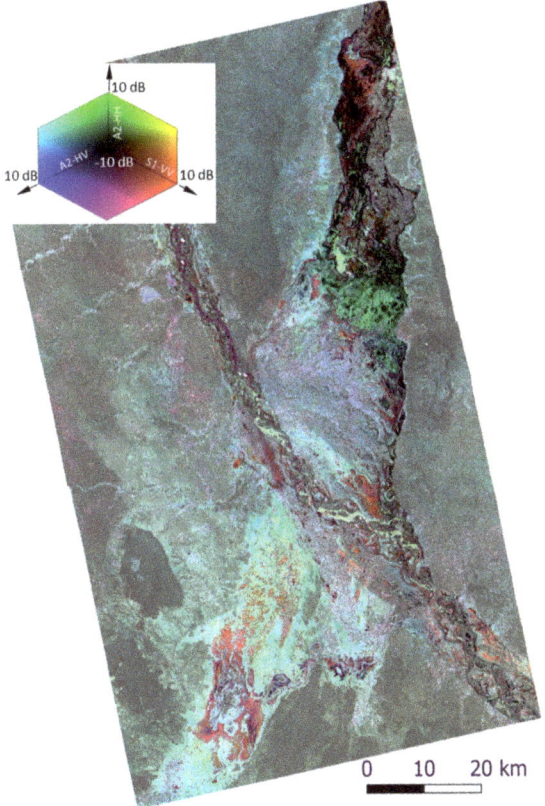

Figure 12. Multifrequency RGB (red-green-blue) intensity ratio combination: red—S1-VV change, green—A2-HH change, blue—A2-HV change. Input values are saturated at ±10 dB, as illustrated by the inset color cube.

In contrast, bright colors denote increase (positive change) in backscatter levels, hinting to the possible presence of flooded vegetation or other structures with double bounce behavior. For instance, on bright red areas, C-VV backscatter increases, while both L-HH and L-HV decrease. These may correspond to short vegetation, such as shrubs, herbaceous vegetation, or cropland, where shorter wavelengths experience double bounce by the interaction of stems/leaves and the underlying water surface, more than longer wavelengths. Vice versa, bright green and cyan areas denote increase in both L-HH and L-HV channels, respectively, with decrease or no change in C-VV. These may indicate the presence of flooded forest or wetlands, where tree trunks or other thick structures contribute to backscattering of longer wavelengths.

These considerations seem confirmed by comparing the RGB-backscatter ratio map with the indicative CORINE land cover map. In the following, we discuss each of the detail areas highlighted by white rectangles in Figure 10 integrating information from multiple sources. The discussion is presented for each of the clusters corresponding to flooded areas.

6. Discussion

Cluster 3 is composed of pixels falling mostly (about 40%) in the class Deciduous Broadleaf Open Forest (Figure 11), with lower percentages of Deciduous Broadleaf Closed Forest and Cropland (both about 17%). Its center backscatter exhibits consistent decrease in L-band data (about 7 dB), but a negligible decrease (about 1 dB) in C-band. The example in Figure 13 shows a spatial localization of this cluster at the border of a thickly vegetated area. The strong decrease in L-band backscatter indicates the presence of open water, while the C-band response seems to be that of a vegetated canopy. The detail inset on the bottom-left of the figure shows a forest glade, which corresponds to pixels in this cluster. Most of these areas seem in fact to correspond to clearings in the forest canopy, exposing the underlying surface, likely covered by low vegetation. During the flood, such clearings can be penetrated by L-band radiation, thus causing the darkening, but not by C-band shorter wavelengths.

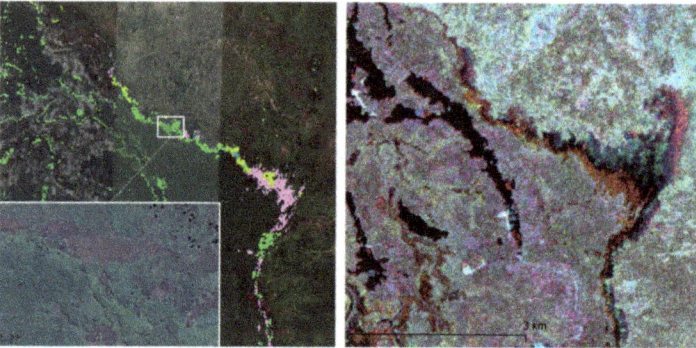

Figure 13. Detail area (**a**) in Figure 10. Left: pixels corresponding to cluster 3, colored according to the underlying CORINE class; inset shows a particular area corresponding to a forest glade; background from Google Earth^{TM} imagery. Right: same area, extracted from the RGB change image in Figure 12.

Figure 14 shows, at the top, pixels corresponding to clusters 4 and 5, colored according to the CORINE class legend. These clusters exhibit similar compositions in terms of CORINE land cover classes on the ground, with 36 to 40% of Deciduous Broadleaf Open Forest, 24 to 32% of Cropland, and 18–20% of Herbaceous Vegetation (Figure 11). The cluster center coordinates in the 6-dimensional backscatter space correspond to roughly constant levels in C band, around −7 to −8 dB, and a slight increase of about 3 dB in L band. The spatial support of this cluster is rather large, corresponding to a relatively large variance of its components. Nevertheless, it includes locally some spatially compact areas, such as those shown in Figure 14. The corresponding RGB image at the bottom of the figure shows the correspondence of the cluster pixels with bright blue/white areas, corresponding to rather strong backscatter increase in both HH and HV L-band, or even in all three channels. The detail inset highlights the different texture in the forest canopy cover corresponding to the cluster pixels, indicating the likely occurrence of more open forest stands, allowing double-bounce phenomena when floodwater is present below the canopy.

Figure 14. Detail area (**b**) in Figure 10. Top: pixels corresponding to clusters 4 and 5, colored according to the underlying CORINE class; inset shows a detail area (see text for explanation); background from Google Earth^{TM} imagery. Bottom: same area, extracted from the RGB change image in Figure 12.

Cluster 6 pixels mostly fall within the Herbaceous Vegetation CORINE category (~54%), besides a lower percentage of the ubiquitous Deciduous Broadleaf Open Forest class (~32%) and negligible other classes (Figure 11). Its centroid corresponds to high backscatter values in C band, with an increase of almost 5 dB from ca. −8 dB to about −3 dB, while in L band backscatter stays quite constant at very low values, −19 dB in HH to −28 dB in HV polarization (Figure 10). An example of a quite compact area corresponding to this cluster is shown in Figure 15 (area c). The predominant color is in fact yellow, corresponding to Herbaceous Vegetation with smaller areas of Deciduous Broadleaf Open Forest in green. In the corresponding RGB composite change image, at the bottom of the figure, the area is mostly colored in light red, indicating in fact increase of C-band levels and no change in the other two channels. This behavior can be explained, in C band, with the double bounce interaction of thin plant stems and branches, typical of herbaceous vegetation, with underlying water, while this effect is not present in L band, due to the longer wavelength. This area appears completely dark in the S1 image acquired in January and it is classified there as "open" water. It is worth noticing that January was the peak of the flood and that the herbaceous vegetation was then likely completely submerged.

Cluster 7 falls on comparable percentages of Deciduous Broadleaf Open Forest (~22%), Herbaceous Wetland (~24%), Cropland (~20%), and Herbaceous Vegetation (~22%) (Figure 11). Its centroid exhibits a strong, −10 dB decrease in C band, while rather constant, low values in L band (Figure 10). Figure 16 shows a representative area with pixels in this cluster, as usual colored as in the CORINE color legend, showing a rather random mix of classes belonging to the above mentioned four, including a wide, compact strip of cropland (in pink), as well as large patches of wetlands (in blue/grey color). Most of the area has green color in the RGB backscatter change composite, at the bottom of the figure, with brighter shades likely corresponding to strips of vegetation along water channels, while darker tones characterize areas farther from the water courses. This in fact corresponds to strong decrease in C-band backscatter, with lower to no decrease in L-band, slightly more pronounced in HV (blue channel) than in HH (green channel) polarization. The likely interpretation of this cluster is of areas normally covered by water (such as wetlands), but which, in correspondence with the investigated event, witness an increase in water levels, overcoming the height of some short and sparse vegetation, which render the surface more specular in C band, while the effect is negligible in L band, giving almost no change in this band.

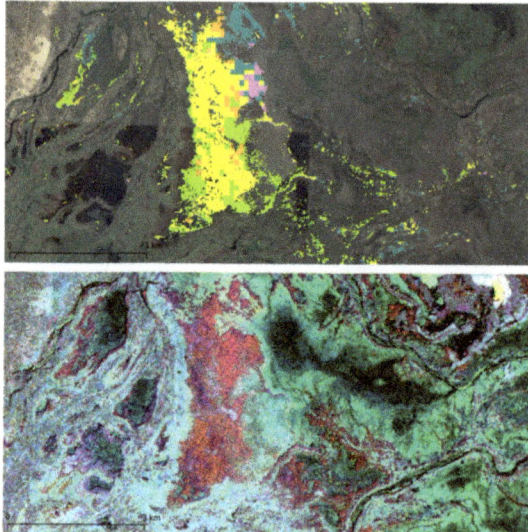

Figure 15. Detail area (**c**) in Figure 10. Top: pixels corresponding to cluster 6, colored according to the underlying CORINE class; background from Google Earth™ imagery. Bottom: same area, extracted from the RGB change image in Figure 12.

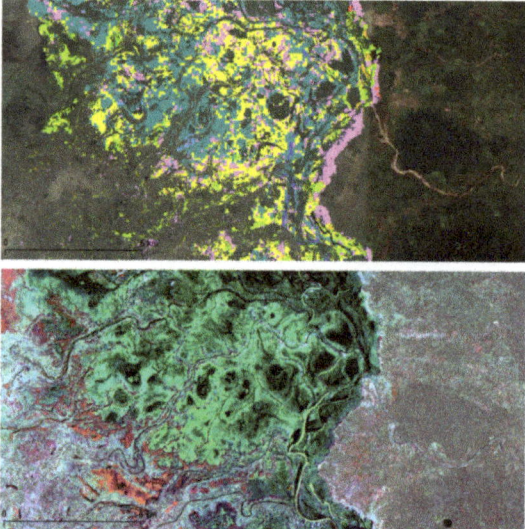

Figure 16. Detail area (**d**) in Figure 10. Top: pixels corresponding to cluster 7, colored according to the underlying CORINE class; background from Google Earth™ imagery. Bottom: same area, extracted from the RGB change image in Figure 12.

Cluster 8 has an even more variegated CORINE class composition, including Deciduous Broadleaf Open Forest (~27%), and then Herbaceous Wetland, Herbaceous Vegetation, and Cropland, each not exceeding 23% (Figure 11). This cluster seems to correspond quite precisely to areas with open water due to the flood, causing a generalized, strong backscatter decrease in both C and L band. Its position in the three plots in Figure 10 is indeed in the bottom-right quadrant, with decrease of about 10 dB in all three cases. Sample cluster areas, represented with the usual CORINE class color map in the left panel of Figure 17, correspond quite precisely with dark areas in the RGB composite on the right, indicating in fact strong generalized backscatter decrease.

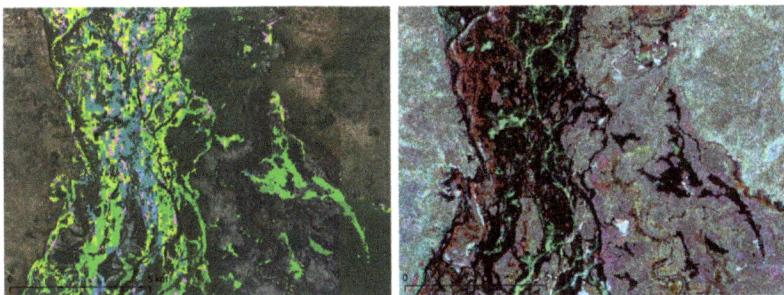

Figure 17. Detail area (**e**) in Figure 10. Left: pixels corresponding to cluster 8, colored according to the underlying CORINE class; background from Google Earth™ imagery. Right: same area, extracted from the RGB change image in Figure 12.

Cluster 9 covers mostly permanent water areas, which indeed form about 48% of its content as CORINE class (Figure 11). The cluster centroid is placed almost exactly on the main diagonal in all three plots in Figure 10, indicating constantly low backscatter values in all channels. Lower, but non-negligible percentages of forested classes are also covered. This can be understood by looking at Figure 18: in the left panel, we show a sample area with the cluster pixels colored in the usual CORINE color code. It can be noticed that most of the pixel correspond to the river course path, in blue color, but this is flanked by a thin strip of pixels flagged as forest (green) or cropland (pink). This non-perfect overlap of SAR derived and CORINE classes can be due to CORINE classification errors, to actual changed conditions on the ground (e.g., seasonal enlargements of the river bed), or both. An even more interesting area is the one shown in the right panel of the same figure. Here, pixels belonging to the SAR-identified permanent water cluster correspond in the CORINE map to vegetated areas, including a narrow water channel (probably too narrow to be "seen" by the coarse-resolution PROBA-V optical sensors used to produce the CORINE map), and several large ponds, equally not identified as permanent waters in the CORINE map, probably because of changed environmental conditions, among the period of PROBA-V imagery (spanning several acquisitions in the interval 2015–2018) and those of the SAR data takes (concentrated in early 2015).

Finally, we focus on cluster 12, which has a CORINE pixel composition not dissimilar from other cluster such as 5, with a high percentage of Deciduous Broadleaf Open Forest (>44%), and lower percentages of cropland and Herbaceous Vegetation (22 and 16%, respectively, Figure 11). Its centroid is roughly on the main diagonal of the C-band plot in Figure 10, while exhibiting decrease of 5–6 dB in L-band levels. Its ground cover is however heterogeneous, as its variances in all the six channels are relatively high. In fact, areas with pixels falling within this cluster may have different behaviors. We choose to show the area in Figure 19 (corresponding to window h in Figure 10), which involves a rather large patch of forest and herbaceous vegetation areas as per the CORINE map, corresponding to bright red color in the RGB composite in the bottom map. This corresponds to very strong positive change in C-band, thus standing for double-bounce increase due to water flooding of (low) vegetation, with equally strong decrease of L-band backscatter levels, thus appearing as open water at longer wavelengths. The most likely interpretation here is of a vegetation with small stems which enhance backscatter in C band, while being specular in L band. Another perhaps more likely phenomenon is the change in water levels in the two dates corresponding to C- and L-band acquisitions, with low levels in the C-band image, thus leaving vegetation sticking out from the water, while higher levels in the L-band acquisition, covering completely the short vegetation cover and causing specular reflection. This situation appears as the predominant one on the forest clearing shown on the Google imagery in the inset detail map.

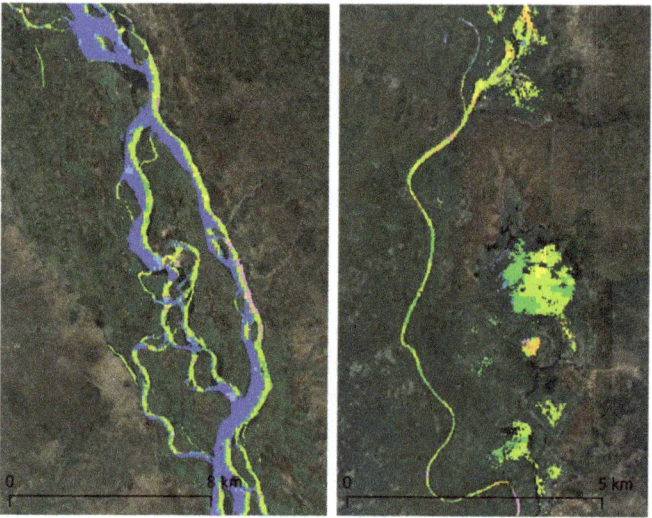

Figure 18. Left: Detail area (**f**) in Figure 10. Right: Detail area (**g**) in Figure 10. Pixels corresponding to cluster 9 are colored according to the underlying CORINE class; background from Google Earth^{TM} imagery.

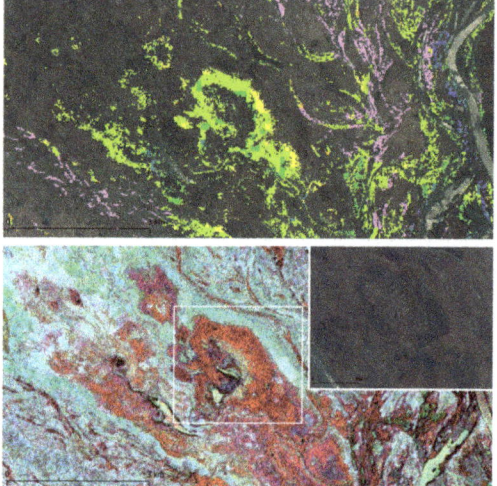

Figure 19. Detail area (**h**) in Figure 10. Top: pixels corresponding to cluster 12, colored according to the underlying CORINE class; background from Google Earth^{TM} imagery. Bottom: same area, extracted from the RGB change image in Figure 12. Inset shows a forest clearing corresponding to the reddish area in the SAR change color composite.

The above-described clusters can be finally cast into a map of flooded areas for the February 2015 event, highlighting the different types of microwave interaction, also according to the CORINE land cover interpretation. The map is shown in Figure 20. A heuristic mask, based on the apparent limits of the flooded areas derived from the preliminary maps of open water derived from either sensor, has been used to spatially constrain the occurrence of areas corresponding to clusters 4 and 5, which basically isolate the flooded forest stands, exhibiting strong double bounce phenomena in L-band. The map shows a rather complex texture of open water, partially and fully submerged vegetated areas, permanently flooded forests and wetlands, and forest openings at the borders of thicker stands where neither C nor L band can penetrate.

The multi-sensor map in Figure 20, together with the multi-temporal maps in Figures 2 and 4, can be regarded as the main contributions of this study to the body of knowledge about remote-sensing-based flood monitoring. We remark the following final points.

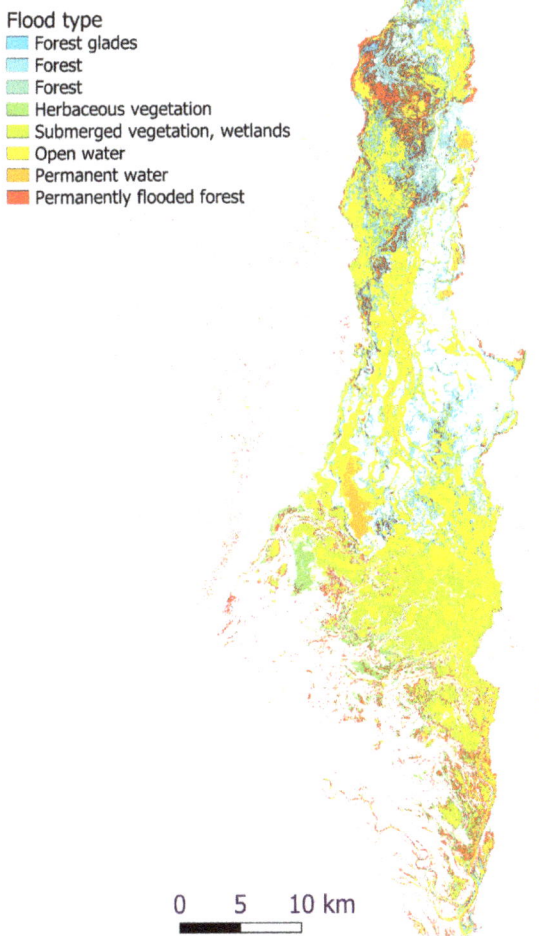

Figure 20. Flood map resulting from the integration of the multi-frequency information. Legend reports the interpretation of the cluster typologies, colored according to the scheme as in Figure 10.

Dense time series of homogeneous acquisitions from the same SAR sensor are very helpful in following the drying dynamics of a given flood event, providing information which can be exploited by hydrologists or environmental scientists to study, e.g., climate change impacts on extreme events dynamics. Currently, the only sensor able to provide such dense time series is Sentinel-1, although other missions such as the X-band Italian constellation COSMO-SkyMed are planned to increase their temporal acquisition schedule. L-band sensors such as ALOS/PALSAR 2, as shown in this work, currently do not allow more than a single co-event image for typical flood durations Multi-frequency integration is extremely useful to recognize different flooded surfaces, especially on vegetated areas, provided that simultaneous acquisitions are available. These at present are quite rare and fortuitous combinations.

Another important part of this study is land cover information, provided by the CORINE database. Although not perfectly "tuned" for these specific applications, both in terms of acquisition times and resolution, it provided invaluable insight into the nature of classes of backscattering conditions on the ground. We remark that, as for instance in the case of permanent waters, SAR-derived information could be exploited in return to update such databases.

Finally, we still remark the absence of any known independent ground truth information for this event. As underlined earlier, this is not an unusual condition for flood events, especially in less developed countries. Nevertheless, application of automated data analysis tools, together with a good deal of heuristic inference, based on indirect evidence from high-resolution optical imagery taken at various times (Google Earth^{TM}), helped in devising convincing map products. We believe further studies should be directed to the automation of such heuristic inference, e.g., through machine learning techniques.

7. Conclusions

Flood monitoring on thickly vegetated, remote areas is important for damage assessment, as well as for studying the response and evolution of inundation phenomena in tropical countries. However, identification of water on the ground, as well as monitoring the event evolution can be challenging, due to different ground cover causing heterogenous response of the terrain surface to the presence of floodwater depending on the type of terrain and the thickness of the vegetation canopy.

In this work, we show an experiment on the integration of multi-temporal, multi-sensor, and multi-polarization SAR data with CORINE land cover information to infer consistent information about a flood phenomenon occurred in early 2015 on the African Shire River basin, in Mozambique.

We first extract information about the temporal evolution of open water flooded areas, through a K-means cluster analysis of the pixels of six Sentinel-1 images acquired at short time intervals during the event. The analysis evidenced the presence of floodwater extending over areas covered by herbaceous vegetation and cropland in the first phases of the flood, followed by a progressive shrinking of the inundated area, with final coverage of wetlands and a few forested areas. Quantification of areas affecting each of the CORINE land cover classes confirms the initial preponderance of flooded herbaceous land cover, which appears to dry faster than wetlands and forests. A similar temporal analysis performed on the two L-band images, one pre- and one co-event, highlighted significant differences in the extent and location of open water with respect to those detected in C band.

Exploiting a time interval in which no significant change is observed in the preceding temporal analysis, a multi-frequency, multi-temporal dataset including pre- and co-event imagery from Sentinel-1 (C band) and ALOS 2 (L band) sensors is then built and analyzed, again through K-means pixel clustering and comparison with CORINE land cover classes and Google imagery. The results highlight the likely presence of floodwater on different types of terrain cover, giving rise to different decrease and increase of backscatter levels in the different bands and polarizations. In particular, this allowed to determine the presence of several areas in which water is present underneath various types of vegetation causing double bounce phenomena of various intensity. A multi-sensor flood map highlighting the different interactions of floodwaters with vegetation according to the used radiation wavelengths has been finally obtained.

This kind of studies are expected to assume increasing importance as the availability of multi-frequency data from SAR satellite constellations will increase in the future. Indeed, to augment its acquisition frequency and to fill critical information gaps in the monitoring of geo-hazards at global scale by extending ground motion information to vegetated areas and by improving flood mapping, especially below vegetation, the Copernicus program is planning to include an L-band Sentinel-1-like satellite, namely ROSE-L, which is part of the six high-priority candidate missions being studied [36].

Author Contributions: All authors contributed equally to paper conceptualization, methodology setup, data analysis and validation; M.Z. and M.C.—original draft preparation; A.R. and A.D.—review and editing. All authors have read and agreed to the published version of the manuscript.

Funding: The contribution of A.R. and A.D. was supported by the Italian Ministry of Education, University and Research, in the framework of the project OT4CLIMA, D.D. 2261–6.9.2018, PON R&I 2014–2020. M. Chini contribution was supported by the Luxembourg National Research Fund through the MOSQUITO project (Grant CORE C15/SR/10380137).

Acknowledgments: The ALOS-2 PALSAR-2 data used in this study are owned by the Japan Aerospace Exploration Agency (JAXA) and were provided through the JAXA's ALOS-2 research program (RA4, PI No. 1255).

Conflicts of Interest: The authors declare no conflict of interest.

References

1. Refice, A.; D'Addabbo, A.; Capolongo, D. (Eds.) *Flood Monitoring through Remote Sensing*; Springer Remote Sensing/Photogrammetry; Springer International Publishing: Cham, Switzerland, 2018; ISBN 978-3-319-63958-1.
2. Schumann, G.J.-P.; Bates, P.D.; Neal, J.C.; Andreadis, K.M. Measuring and Mapping Flood Processes. In *Hydro-Meteorological Hazards, Risks and Disasters*; Elsevier: Amsterdam, The Netherlands, 2015; pp. 35–64, ISBN 9780123964700.
3. Rahman, M.S.; Di, L. A Systematic Review on Case Studies of Remote-Sensing-Based Flood Crop Loss Assessment. *Agriculture* **2020**, *10*, 131. [CrossRef]
4. Refice, A.; Capolongo, D.; Pasquariello, G.; D'Addabbo, A.; Bovenga, F.; Nutricato, R.; Lovergine, F.P.; Pietranera, L. SAR and InSAR for Flood Monitoring: Examples with COSMO-SkyMed Data. *IEEE J. Sel. Top. Appl. Earth Obs. Remote Sens.* **2014**, *7*, 2711–2722. [CrossRef]
5. Pulvirenti, L.; Pierdicca, N.; Chini, M.; Guerriero, L. Monitoring flood evolution in vegetated areas using cosmo-skymed data: The tuscany 2009 case study. *IEEE J. Sel. Top. Appl. Earth Obs. Remote Sens.* **2013**, *6*, 1807–1816. [CrossRef]
6. Pierdicca, N.; Pulvirenti, L.; Chini, M.; Guerriero, L.; Candela, L. Observing floods from space: Experience gained from COSMO-SkyMed observations. *Acta Astronautica* **2013**, *84*, 122–133. [CrossRef]
7. Schlaffer, S.; Matgen, P.; Hollaus, M.; Wagner, W. Flood detection from multi-temporal SAR data using harmonic analysis and change detection. *Int. J. Appl. Earth Obs. Geoinf.* **2015**, *38*, 15–24. [CrossRef]
8. Capolongo, D.; Refice, A.; Bocchiola, D.; D'Addabbo, A.; Vouvalidis, K.; Soncini, A.; Zingaro, M.; Bovenga, F.; Stamatopoulos, L. Coupling multitemporal remote sensing with geomorphology and hydrological modeling for post flood recovery in the Strymonas dammed river basin (Greece). *Sci. Total Environ.* **2019**, *651*, 1958–1968. [CrossRef]
9. Pulvirenti, L.; Chini, M.; Pierdicca, N.; Guerriero, L.; Ferrazzoli, P. Flood monitoring using multi-temporal COSMO-SkyMed data: Image segmentation and signature interpretation. *Remote Sens. Environ.* **2011**, *115*, 990–1002. [CrossRef]
10. Chini, M.; Hostache, R.; Giustarini, L.; Matgen, P. A hierarchical split-based approach for parametric thresholding of SAR images: Flood inundation as a test case. *IEEE Trans. Geosci. Remote Sens.* **2017**, *55*, 6975–6988. [CrossRef]
11. Hess, L.L.; Melack, J.M.; Simonett, D.S. Radar detection of flooding beneath the forest canopy: A review. *Int. J. Remote Sens.* **1990**, *11*, 1313–1325. [CrossRef]

12. Pierdicca, N.; Pulvirenti, L.; Chini, M. Flood Mapping in Vegetated and Urban Areas and Other Challenges: Models and Methods. In *Flood Monitoring through Remote Sensing*; Springer: Cham, Switzerland, 2018; pp. 135–179.
13. Wang, Y.; Imhoff, M.L. Simulated and observed L-HH radar backscatter from tropical mangrove forests. *Int. J. Remote Sens.* **1993**, *14*, 2819–2828. [CrossRef]
14. Dabrowska-Zielinska, K.; Budzynska, M.; Tomaszewska, M.; Bartold, M.; Gatkowska, M.; Malek, I.; Turlej, K.; Napiorkowska, M. Monitoring wetlands ecosystems using ALOS PALSAR (L-Band, HV) supplemented by optical data: A case study of Biebrza Wetlands in Northeast Poland. *Remote Sens.* **2014**, *6*, 1605–1633. [CrossRef]
15. Dabboor, M.; Brisco, B. Wetland Monitoring and Mapping Using Synthetic Aperture Radar. In *Wetlands*; Intech Open: London, UK, 2018; pp. 1–26.
16. Manavalan, R.; Rao, Y.S.; Krishna Mohan, B. Comparative flood area analysis of C-band VH, VV, and L-band HH polarizations SAR data. *Int. J. Remote Sens.* **2017**, *38*, 4645–4654. [CrossRef]
17. Pierdicca, N.; Pulvirenti, L.; Boni, G.; Squicciarino, G.; Chini, M. Mapping Flooded Vegetation Using COSMO-SkyMed: Comparison with Polarimetric and Optical Data Over Rice Fields. *IEEE J. Sel. Top. Appl. Earth Obs. Remote Sens.* **2017**, *10*, 2650–2662. [CrossRef]
18. D'Addabbo, A.; Refice, A.; Pasquariello, G.; Lovergine, F.P.; Capolongo, D.; Manfreda, S. A Bayesian Network for Flood Detection Combining SAR Imagery and Ancillary Data. *IEEE Trans. Geosci. Remote Sens.* **2016**, *54*, 3612–3625. [CrossRef]
19. Hong, S.-H.; Wdowinski, S. Double-Bounce Component in Cross-Polarimetric SAR From a New Scattering Target Decomposition. *IEEE Trans. Geosci. Remote Sens.* **2014**, *52*, 3039–3051. [CrossRef]
20. Schlaffer, S.; Chini, M.; Dettmering, D.; Wagner, W. Mapping Wetlands in Zambia Using Seasonal Backscatter Signatures Derived from ENVISAT ASAR Time Series. *Remote Sens.* **2016**, *8*, 402. [CrossRef]
21. Beilfuss, R.D.; dos Santos, D. *Patterns of Hydrological Change in the Zambezi Delta, Mozambique*; Working Paper #2 Zambezi Basin Crane and Wetland Conservation Program; International Crane Foundation: Baraboo, WI, USA, 2001.
22. Nharo, T. Modeling Floods in the Middle Zambezi Basin Using Remote Sensing and Hydrological Modeling Techniques. Master's Thesis, University of Zimbabwe, Harare, Zimbabwe, 2016.
23. Singh, A. *Historical Analysis of Change in Rainfall, Flow Regime and River Morphology Case Study: Zambezi Delta*; UNESCO-IHE Institute for Water Education: Delft, The Netherlands, 2017.
24. Beilfuss, R.D.; Moore, D.; Bento, C.; Dutton, P. *Patterns of Vegetation Change in the Zambezi Delta, Mozambique*; Working Paper #3 Zambezi Basin Crane and Wetland Conservation Program; International Crane Foundation: Baraboo, WI, USA, 2001.
25. Davies, B.R.; Beilfuss, R.D.; Thoms, M.C. Cahora Bassa retrospective, 1974–1997: Effects of flow regulation on the Lower Zambezi River. *SIL Proc.* **2000**, *27*, 2149–2157. [CrossRef]
26. Pasanisi, F.; Tebano, C.; Zarlenga, F. *Indagine Morfologica su un Tratto del Fiume Zambesi, nella Repubblica del Mozambico. Misure Batimetriche ed Analisi Dei Dati*; Technical Report No. RT/2015/14/ENEA; ENEA: Rome, Italy, 2015.
27. Ronco, P.; Fasolato, G.; Nones, M.; Di Silvio, G. Morphological effects of damming on lower Zambezi River. *Geomorphology* **2010**, *115*, 43–55. [CrossRef]
28. Shela, O. Management of shared river basins: The case of the Zambezi River. *Water Policy* **2000**, *2*, 65–81. [CrossRef]
29. Lee, J.-S.; Ainsworth, T.L.; Wang, Y.; Chen, K.-S. Polarimetric SAR Speckle Filtering and the Extended Sigma Filter. *IEEE Trans. Geosci. Remote Sens.* **2015**, *53*, 1150–1160. [CrossRef]
30. ESA. *Land Cover CCI Product User Guide Version 2.0*; ESA: Paris, France, 2017.
31. Refice, A.; D'Addabbo, A.; Lovergine, F.P.; Tijani, K.; Morea, A.; Nutricato, R.; Bovenga, F.; Nitti, D.O. Monitoring Flood Extent and Area Through Multisensor, Multi-temporal Remote Sensing: The Strymonas (Greece) River Flood. In *Flood Monitoring through Remote Sensing*; Springer International Publishing: Cham, Switzerland, 2018; pp. 101–113.
32. MacQueen, J. Some Methods for Classification and Analysis of Multivariate Observations. In *Proceedings of the 5th Berkeley Symposium on Mathematical Statistics and Probability*; University of California Press: Berkeley/Los Angeles, CA, USA, 1967; Volume 1, pp. 281–297.

33. D'Addabbo, A.; Refice, A.; Lovergine, F.P.; Pasquariello, G. DAFNE: A Matlab toolbox for Bayesian multi-source remote sensing and ancillary data fusion, with application to flood mapping. *Comput. Geosci.* **2018**, *112*, 64–75. [CrossRef]
34. Chini, M.; Chiancone, A.; Stramondo, S. Scale Object Selection (SOS) through a hierarchical segmentation by a multi-spectral per-pixel classification. *Pattern Recognit. Lett.* **2014**, *49*, 214–223. [CrossRef]
35. Bhattacharya, A. On a measure of divergence between two multinomial populations. *Indian J. Stat.* **1946**, *7*, 401–406.
36. Pierdicca, N.; Davidson, M.; Chini, M.; Dierking, W.; Djavidnia, S.; Haarpaintner, J.; Hajduch, G.; Laurin, G.V.; Lavalle, M.; López-Martínez, C.; et al. The Copernicus L-band SAR mission ROSE-L (Radar Observing System for Europe). In *Active and Passive Microwave Remote Sensing for Environmental Monitoring III*; Notarnicola, C., Pierdicca, N., Bovenga, F., Santi, E., Eds.; SPIE: Washington, DC, USA, 2019; Volume 11154, p. 13.

© 2020 by the authors. Licensee MDPI, Basel, Switzerland. This article is an open access article distributed under the terms and conditions of the Creative Commons Attribution (CC BY) license (http://creativecommons.org/licenses/by/4.0/).

MDPI
St. Alban-Anlage 66
4052 Basel
Switzerland
Tel. +41 61 683 77 34
Fax +41 61 302 89 18
www.mdpi.com

Water Editorial Office
E-mail: water@mdpi.com
www.mdpi.com/journal/water

www.ingramcontent.com/pod-product-compliance
Lightning Source LLC
LaVergne TN
LVHW070620100526
838202LV00012B/688